Fundamentals of Airplane Flight Mechanics

David G. Hull

Fundamentals of Airplane Flight Mechanics

With 125 Figures and 25 Tables

 Springer

David G. Hull
The University of Texas at Austin
Aerospace Engineering and Engineering Mechanics
1, University Station, C0600
Austin, TX 78712-0235
USA
e-mail: dghull@mail.utexas.edu

ISBN   978-3-642-07987-0   e-ISBN   978-3-540-46573-7

Springer is a part of Springer Science+Business Media.

springer.com

© Springer-Verlag Berlin Heidelberg 2010

Cover design: eStudio, Calamar, Girona, Spain

David G. Hull

Fundamentals of Airplane
Flight Mechanics

With 125 Figures and 25 Tables

 Springer

David G. Hull
The University of Texas at Austin
Aerospace Engineering and Engineering Mechanics
1, University Station, C0600
Austin, TX 78712-0235
USA
e-mail: dghull@mail.utexas.edu

ISBN   978-3-642-07987-0   e-ISBN   978-3-540-46573-7

Cover design: eStudio, Calamar, Girona, Spain

Dedicated to

Angelo Miele

who instilled in me his love for flight mechanics.

Preface

Flight mechanics is the application of Newton's laws (F=ma and M=Iα) to the study of vehicle trajectories (performance), stability, and aerodynamic control. There are two basic problems in airplane flight mechanics: (1) given an airplane what are its performance, stability, and control characteristics? and (2) given performance, stability, and control characteristics, what is the airplane? The latter is called airplane sizing and is based on the definition of a standard mission profile. For commercial airplanes including business jets, the mission legs are take-off, climb, cruise, descent, and landing. For a military airplane additional legs are the supersonic dash, fuel for air combat, and specific excess power. This text is concerned with the first problem, but its organization is motivated by the structure of the second problem. Trajectory analysis is used to derive formulas and/or algorithms for computing the distance, time, and fuel along each mission leg. In the sizing process, all airplanes are required to be statically stable. While dynamic stability is not required in the sizing process, the linearized equations of motion are used in the design of automatic flight control systems.

This text is primarily concerned with analytical solutions of airplane flight mechanics problems. Its design is based on the precepts that there is only one semester available for the teaching of airplane flight mechanics and that it is important to cover both trajectory analysis and stability and control in this course. To include the fundamentals of both topics, the text is limited mainly to flight in a vertical plane. This is not very restrictive because, with the exception of turns, the basic trajectory segments of both mission profiles and the stability calculations are in the vertical plane. At the University of Texas at Austin, this course is preceded by courses on low-speed aerodynamics and linear system theory. It is followed by a course on automatic control.

The trajectory analysis portion of this text is patterned after Miele's flight mechanics text in terms of the nomenclature and the equations of motion approach. The aerodynamics prediction algorithms have been taken from an early version of the NASA-developed business jet sizing code called the General Aviation Synthesis Program or GASP. An important part of trajectory analysis is trajectory optimization. Ordinarily, trajectory optimization is a complicated affair involving optimal control theory (calculus of variations) and/or the use of numerical optimization techniques. However, for the standard mission legs, the optimization problems are quite simple in nature. Their solution can be obtained through the use of basic calculus.

The nomenclature of the stability and control part of the text is based on the writings of Roskam. Aerodynamic prediction follows that of the USAF Stability and Control Datcom. It is important to be able to list relatively simple formulas for predicting aerodynamic quantities and to be able to carry out these calculations throughout performance, stability, and control. Hence, it is assumed that the airplanes have straight, tapered, swept wing planforms.

Flight mechanics is a discipline. As such, it has equations of motion, acceptable approximations, and solution techniques for the approximate equations of motion. Once an analytical solution has been obtained, it is important to calculate some numbers to compare the answer with the assumptions used to derive it and to acquaint students with the sizes of the numbers. The Subsonic Business Jet (SBJ) defined in App. A is used for these calculations.

The text is divided into two parts: trajectory analysis and stability and control. To study trajectories, the force equations (F=ma) are uncoupled from the moment equations (M=Iα) by assuming that the airplane is not rotating and that control surface deflections do not change lift and drag. The resulting equations are referred to as the 3DOF model, and their investigation is called trajectory analysis. To study stability and control, both F=ma and M=Iα are needed, and the resulting equations are referred to as the 6DOF model. An overview of airplane flight mechanics is presented in Chap. 1.

Part I: Trajectory Analysis. This part begins in Chap. 2 with the derivation of the 3DOF equations of motion for flight in a vertical plane over a flat earth and their discussion for nonsteady flight and quasi-steady flight. Next in Chap. 3, the atmosphere (standard and exponential) is discussed, and an algorithm is presented for computing lift and drag of a subsonic airplane. The engines are assumed to be given, and the thrust and specific fuel consumption are discussed for a subsonic turbojet and turbofan. Next, the quasi-steady flight problems of cruise and climb are analyzed in Chap. 4 for an arbitrary airplane and in Chap. 5 for an ideal subsonic airplane. In Chap. 6, an algorithm is presented for calculating the aerodynamics of high-lift devices, and the nonsteady flight problems of take-off and landing are discussed. Finally, the nonsteady flight problems of energy climbs, specific excess power, energy-maneuverability, and horizontal turns are studied in Chap. 7.

Part II: Stability and Control. This part of the text contains static stability and control and dynamic stability and control. It is begun in Chap. 8 with the 6DOF model in wind axes. Following the discussion of the equations of motion, formulas are presented for calculating the aerodynamics of a subsonic airplane including the lift, the pitching moment, and the drag. Chap. 9 deals with static stability and control. Trim conditions and static stability are investigated for steady cruise, climb, and descent along with the effects of center of gravity position. A simple control system is analyzed to introduce the concepts of hinge moment, stick force, stick force gradient, and handling qualities. Trim tabs and the effect of free elevator on stability are discussed. Next, trim conditions are determined for a nonsteady pull-up, and lateral-directional stability and control are discussed briefly. In Chap. 10, the 6DOF equations of motion are developed first in regular body axes and second in stability axes for use in the investigation of dynamic stability and control. In Chap. 11, the equations of motion are linearized about a steady reference path, and the stability and response of an airplane to a control or gust input is considered. Finally, the effect of center of gravity position is examined, and dynamic lateral-direction stability and control is discussed descriptively.

There are three appendices. App. A gives the geometric characteristics of a subsonic business jet, and results for aerodynamic calculations are listed, including both static and dynamic stability and control results. In App. B, the relationship between linearized aerodynamics (stability derivatives) and the aerodynamics of Chap. 8 is established. Finally, App. C reviews the elements of linear system theory which are needed for dynamic stability and control studies.

While a number of students has worked on this text, the author is particularly indebted to David E. Salguero. His work on converting GASP into an educational tool called BIZJET has formed the basis of a lot of this text.

David G. Hull
Austin, Texas

Table of Contents

Chapter 1

Introduction to Airplane Flight Mechanics

Airplane flight mechanics can be divided into five broad areas: trajectory analysis (performance), stability and control, aircraft sizing, simulation, and flight testing. Only the theoretical aspects of trajectory analysis and stability and control are covered in this text. Aircraft sizing and simulation are essentially numerical in nature. Airplane sizing involves an iterative process, and simulation involves the numerical integration of a set of differential equations. They are discussed in this chapter to show how they fit into the overall scheme of things. Flight testing is the experimental part of flight mechanics. It is not discussed here except to say that good theory makes good experiments.

The central theme of this text is the following: Given the three-view drawing with dimensions of a subsonic, jet-powered airplane and the engine data, determine its performance, stability, and control characteristics. To do this, formulas for calculating the aerodynamics are developed.

Most of the material in this text is limited to flight in a vertical plane because the mission profiles for which airplanes are designed are primarily in the vertical plane. This chapter begins with a review of the parts of the airframe and the engines. Then, the derivation of the equations governing the motion of an airplane is discussed. Finally, the major areas of aircraft flight mechanics are described.

1.1   Airframe Anatomy

To begin the introduction, it is useful to review the parts of an airframe and discuss their purposes. Fig. 1.1 is a three-view drawing of a Boeing 727. The body or fuselage of the airplane holds the crew, passengers, and freight. It is carried aloft by the lift provided by the wing and propelled by the thrust produced by jet engines housed in nacelles. This airplane has two body-mounted engines and a body centerline engine whose inlet air comes through an S-duct beginning at the front of the vertical tail. The fuel is carried in tanks located in the wing.

Since a jet transport is designed for efficient high-speed cruise, it is unable to take-off and land from standard-length runways without some configuration change. This is provided partly by leading edge slats and partly by trailing edge flaps. Both devices are used for take-off, with a low trailing edge flap deflection. On landing, a high trailing edge flap deflection is used to increase lift and drag, and brakes, reverse thrust, and speed brakes (spoilers) are used to further reduce landing distance.

A major issue in aircraft design is static stability. An airplane is said to be inherently aerodynamically statically stable if, following a disturbance from a steady flight condition, forces and/or moments develop which tend to reduce the disturbance. Shown in Fig. 1.2 is the body axes system whose origin is at the center of gravity and whose $x_b, y_b,$ and z_b axes are called the roll axis, the pitch axis, and the yaw axis. Static stability about the yaw axis (directional stability) is provided by the vertical stabilizer, whereas the horizontal stabilizer makes the airplane statically stable about the pitch axis (longitudinal stability). Static stability about the roll axis (lateral stability) is provided mainly by wing dihedral which can be seen in the front view in Fig. 1.1.

Also shown in Figs. 1.1 and 1.2 are the control surfaces which are intended to control the rotation rates about the body axes (roll rate P, pitch rate Q, and yaw rate R) by controlling the moments about these axes (roll moment L, pitch moment M, and yaw moment N). The convention for positive moments and rotation rates is to grab an axis with the thumb pointing toward the origin and rotate counterclockwise looking down the axis toward the origin. From the pilot's point of view, a positive moment or rate is roll right, pitch up, and yaw right.

The deflection of a control surface changes the curvature of a wing or tail surface, changes its lift, and changes its moment about the

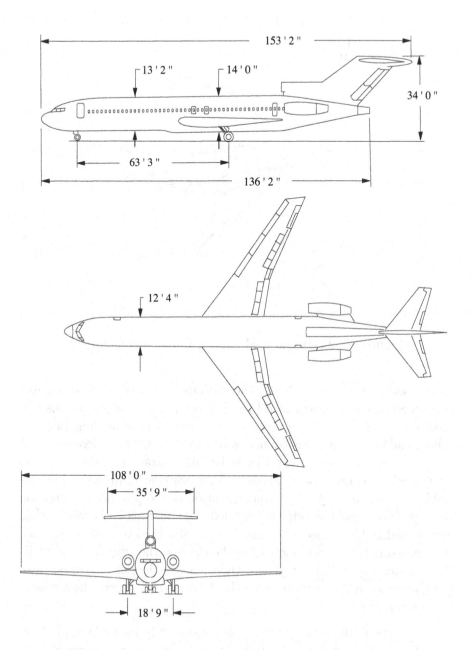

Figure 1.1: Three-View Drawing of a Boeing 727

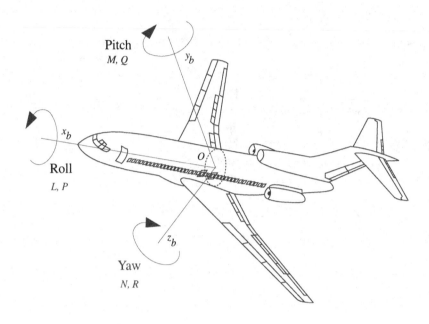

Figure 1.2: Body Axes, Moments, Rates, and Controls

corresponding body axis. Hence, the ailerons (one deflected upward and one deflected downward) control the roll rate; the elevator controls the pitch rate; and the rudder controls the yaw rate. Unlike pitching motion, rolling and yawing motions are not pure. In deflecting the ailerons to roll the airplane, the down-going aileron has more drag than the up-going aileron which causes the airplane to yaw. Similarly, in deflecting the rudder to yaw the airplane, a rolling motion is also produced. Cures for these problems include differentially deflected ailerons and coordinating aileron and rudder deflections. Spoilers are also used to control roll rate by decreasing the lift and increasing the drag on the wing into the turn. Here, a yaw rate is developed into the turn. Spoilers are not used near the ground for roll control because the decreased lift causes the airplane to descend.

The F-16 (lightweight fighter) is statically unstable in pitch at subsonic speeds but becomes statically stable at supersonic speeds because of the change in aerodynamics from subsonic to supersonic speeds. The airplane was designed this way to make the horizontal tail as small as possible and, hence, to make the airplane as light as possible. At subsonic speeds, pitch stability is provided by the automatic flight control system. A rate gyro senses a pitch rate, and if the pitch rate is not commanded by the pilot, the elevator is deflected automatically to zero the pitch rate. All of this happens so rapidly (at the speed of electrons) that the pilot is unaware of these rotations.

1.2 Engine Anatomy

In this section, the various parts of jet engines are discussed. There are two types of jet engines in wide use: the turbojet and the turbofan.

A schematic of a *turbojet* is shown in Fig. 1.3. Air entering the engine passes through the diffuser which slows the air to a desired speed for entering the compressor. The compressor increases the pressure of the air and slows it down more. In the combustion chamber (burner), fuel is added to the air, and the mixture is ignited and burned. Next, the high temperature stream passes through the turbine which extracts enough energy from the stream to run the compressor. Finally, the nozzle increases the speed of the stream before it exits the engine.

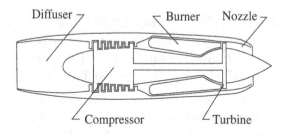

Figure 1.3: Schematic of a Turbojet Engine

The engine cycle is a sequence of assumptions describing how the flow behaves as it passes through the various parts of the engine. Given the engine cycle, it is possible to calculate the thrust (lb) and the fuel flow rate (lb/hr) of the engine. Then, the specific fuel consumption (1/hr) is the ratio of the fuel flow rate to the thrust.

A schematic of a *turbofan* is shown in Fig. 1.4. The turbofan is essentially a turbojet which drives a fan located after the diffuser and before the compressor. The entering air stream is split into a primary part which passes through the turbojet and a secondary part which goes around the turbojet. The split is defined by the bypass ratio, which is the ratio of the air mass flow rate around the turbojet to the air mass flow rate through the turbojet. Usually, the fan is connected to its own turbine by a shaft, and the compressor is connected to its turbine by a hollow shaft which rotates around the fan shaft.

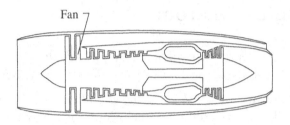

Fan

Figure 1.4: Schematic of a Turbofan Engine

1.3   Equations of Motion

In this text, the term flight mechanics refers to the analysis of airplane motion using Newton's laws. While most aircraft structures are flexible to some extent, the airplane is assumed here to be a rigid body. When fuel is being consumed, the airplane is a variable-mass rigid body.

Newton's laws are valid when written relative to an inertial reference frame, that is, a reference frame which is not accelerating or rotating. If the equations of motion are derived relative to an inertial reference frame and if approximations characteristic of airplane motion are introduced into these equations, the resulting equations are those for flight over a nonrotating flat earth. Hence, for airplane motion, the earth is an approximate inertial reference frame, and this model is called the flat earth model. The use of this physical model leads to a small error in most analyses.

A general derivation of the equations of motion involves the use of a material system involving both solid and fluid particles. The end result is a set of equations giving the motion of the solid part of the airplane subject to aerodynamic, propulsive and gravitational forces. To simplify the derivation of the equations of motion, the correct equations for the forces are assumed to be known. Then, the equations describing the motion of the solid part of the airplane are derived.

The airplane is assumed to have a right-left plane of symmetry with the forces acting at the center of gravity and the moments acting about the center of gravity. Actually, the forces acting on an airplane in fight are due to distributed surface forces and body forces. The surface forces come from the air moving over the airplane and through the propulsion system, while the body forces are due to gravitational effects. Any distributed force (see Fig. 1.5) can be replaced by a concentrated force acting along a specific line of action. Then, to have all forces acting through the same point, the concentrated force can be replaced by the same force acting at the point of interest plus a moment about that point to offset the effect of moving the force. The point usually chosen for this purpose is the center of mass, or equivalently for airplanes the center of gravity, because the equations of motion are the simplest.

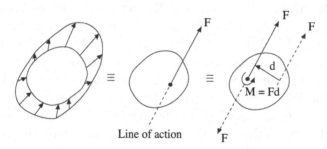

Figure 1.5: Distributed Versus Concentrated Forces

The equations governing the translational and rotational motion of an airplane are the following:

a. Kinematic equations giving the translational position and rotational position relative to the earth reference frame.

b. Dynamic equations relating forces to translational acceleration and moments to rotational acceleration.

c. Equations defining the variable-mass characteristics of the airplane (center of gravity, mass and moments of inertia) versus time.

d. Equations giving the positions of control surfaces and other movable parts of the airplane (landing gear, flaps, wing sweep, etc.) versus time.

These equations are referred to as the six degree of freedom (6DOF) equations of motion. The use of these equations depends on the particular area of flight mechanics being investigated.

1.4   Trajectory Analysis

Most trajectory analysis problems involve small aircraft rotation rates and are studied through the use of the three degree of freedom (3DOF) equations of motion, that is, the translational equations. These equations are uncoupled from the rotational equations by assuming negligible rotation rates and neglecting the effect of control surface deflections on aerodynamic forces. For example, consider an airplane in cruise. To maintain a given speed an elevator deflection is required to make the pitching moment zero. This elevator defection contributes to the lift and the drag of the airplane. By neglecting the contribution of the elevator deflection to the lift and drag (untrimmed aerodynamics), the translational and rotational equations uncouple. Another approach, called trimmed aerodynamics, is to compute the control surface angles required for zero aerodynamic moments and eliminate them from the aerodynamic forces. For example, in cruise the elevator angle for zero aerodynamic pitching moment can be derived and eliminated from the drag and the lift. In this way, the extra aerodynamic force due to control surface deflection can be taken into account.

Trajectory analysis takes one of two forms. First, given an aircraft, find its performance characteristics, that is, maximum speed, ceiling, range, etc. Second, given certain performance characteristics, what is the airplane which produces them. The latter is called aircraft sizing, and the missions used to size commercial and military aircraft are presented here to motivate the discussion of trajectory analysis. The mission or flight profile for sizing a commercial aircraft (including business jets) is shown in Fig. 1.6. It is composed of take-off, climb, cruise, descent, and landing segments, where the descent segment is replaced by an extended cruise because the fuel consumed is approximately the same. In each segment, the distance traveled, the time elapsed, and the fuel consumed must be computed to determine the corresponding quantities for the whole mission. The development of formulas or algorithms for computing these performance quantities is the charge of trajectory analysis. The military mission (Fig. 1.7) adds three performance computations: a constant-altitude acceleration (supersonic dash), constant-altitude turns, and specific excess power (P_S). The low-altitude dash gives the airplane the ability to approach the target within the radar ground clutter, and the speed of the approach gives the airplane the ability to avoid detection until it nears the target. The number of turns is specified to ensure that the airplane has enough fuel for air combat in the neighborhood of the target. Specific excess power is a measure of the ability of the airplane to change its energy, and it is used to ensure that the aircraft being designed has superior maneuver capabilities relative to enemy aircraft protecting the target. Note that, with the exception of the turns, each segment takes place in a plane perpendicular to the surface of the earth (vertical plane). The turns take place in a horizontal plane.

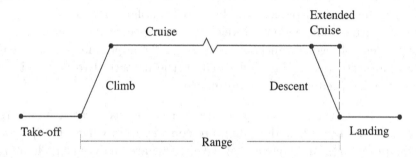

Figure 1.6: Mission for Commercial Aircraft Sizing

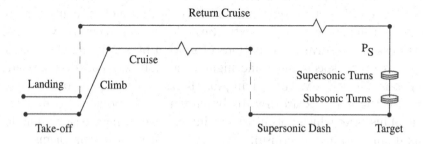

Figure 1.7: Mission for Military Aircraft Sizing

These design missions are the basis for the arrangement of the trajectory analysis portion of this text. In Chap. 2, the equations of motion for flight in a vertical plane over a flat earth are derived, and their solution is discussed. Chap. 3 contains the modeling of the atmosphere, aerodynamics, and propulsion. Both the standard atmosphere and the exponential atmosphere are discussed. An algorithm for predicting the drag polar of a subsonic airplane from a three-view drawing is presented, as is the parabolic drag polar. Engine data is assumed to be available and is presented for a subsonic turbojet and turbofan. Approximate analytical expressions are obtained for thrust and specific fuel consumption.

The mission legs characterized by quasi-steady flight (climb, cruise, and descent) are analyzed in Chap. 4. Algorithms for computing distance, time, and fuel are presented for arbitrary aerodynamics and propulsion, and numerical results are developed for a subsonic business jet. In Chap. 5 approximate analytical results are derived by assuming an ideal subsonic airplane: parabolic drag polar with constant coefficients, thrust independent of velocity, and specific fuel consumption independent of velocity and power setting. The approximate analytical results are intended to be used to check the numerical results and for making quick predictions of performance.

Next, the mission legs characterized by accelerated flight are investigated. Take-off and landing are considered in Chap. 6. Specific excess power, P_S, and constant altitude turns are analyzed in Chap. 7. However, the supersonic dash is not considered because it involves flight through the transonic region.

In general, the airplane is a controllable dynamical system. Hence, the differential equations which govern its motion contain more variables than equations. The extra variables are called control variables. It is possible to solve the equations of motion by specifying the control histories or by specifying some flight condition, say constant altitude and constant velocity, and solving for the controls. On the other hand, because the controls are free to be chosen, it is possible to find the control histories which optimize some index of performance (for example, maximum distance in cruise). Trajectory optimization problems such as these are handled by a mathematical theory known as Calculus of Variations or Optimal Control Theory. While the theory is beyond the scope of this text, many aircraft trajectory optimization problems can be formulated as simple optimization problems whose theory can be derived by simple reasoning.

1.5 Stability and Control

Stability and control studies are concerned with motion of the center of gravity (cg) relative to the ground and motion of the airplane about the cg. Hence, stability and control studies involve the use of the six degree of freedom equations of motion. These studies are divided into two major categories: (a) static stability and control and (b) dynamic stability and control. Because of the nature of the solution process, each of the categories is subdivided into longitudinal motion (pitching motion) and lateral-directional motion (combined rolling and yawing motion). While trajectory analyses are performed in terms of force coefficients with control surface deflections either neglected (untrimmed drag polar) or eliminated (trimmed drag polar), stability and control analyses are in terms of the orientation angles (angle of attack and sideslip angle) and the control surface deflections.

The six degree of freedom model for flight in a vertical plane is presented in Chap. 8. First, the equations of motion are derived in the wind axes system. Second, formulas for calculating subsonic aerodynamics are developed for an airplane with a straight, tapered, swept wing. The aerodynamics associated with lift and pitching moment are shown to be linear in the angle of attack, the elevator angle, the pitch rate, and the angle of attack rate. The aerodynamics associated with drag is shown to be quadratic in angle of attack. Each coefficient in these relationships is a function of Mach number.

Chap. 9 is concerned with static stability and control. Static stability and control for quasi-steady flight is concerned primarily with four topics: trim conditions, static stability, center of gravity effects, and control force and handling qualities. The trim conditions are the orientation angles and control surface deflections required for a particular flight condition. Given a disturbance from a steady flight condition, static stability investigates the tendency of the airplane to reduce the disturbance. This is done by looking at the signs of the forces and moments. Fore and aft limits are imposed on allowable cg locations by maximum allowable control surface deflections and by stability considerations, the aft cg limit being known as the neutral point because it indicates neutral stability. Handling qualities studies are concerned with pilot-related quantities such as control force and how control force changes with flight speed. These quantities are derived from aerodynamic moments about control surface hinge lines. Trim tabs have been introduced to allow the pilot to zero out the control forces associated with a particular flight condition. However, if after trimming the stick force the pilot flies hands-off, the stability characteristics of the airplane are reduced.

To investigate static stability and control for accelerated flight, use is made of a pull-up. Of interest is the elevator angle required to make an n-g turn or pull-up. There is a cg position where the elevator angle per g goes to zero, making the airplane too easy to maneuver. This cg position is called the maneuver point. There is another maneuver point associated with the stick force required to make an n-g pull-up.

While dynamic stability and control studies can be conducted using wind axes, it is the convention to use body axes. Hence, in Chap. 10, the equations of motion are derived in the body axes. The aerodynamics need for body axes is the same as that used in wind axes. A particular set of body axes is called stability axes. The equations of motion are also developed for stability axes.

Dynamic stability and control is concerned with the motion of an airplane following a disturbance such as a wind gust (which changes the speed, the angle of attack and/or the sideslip angle) or a control input. While these studies can and are performed using detailed computer simulations, it is difficult to determine cause and effect. As a consequence, it is desirable to develop an approximate analytical approach. This is done in Chap. 11 by starting with the airplane in a quasi-steady flight condition (given altitude, Mach number, weight, power setting) and introducing a small disturbance. By assuming that the changes in the variables are small, the equations of motion can be linearized about the steady flight condition. This process leads to a system of linear, ordinary differential equations with constant coefficients. As is known from linear system theory, the response of an airplane to a disturbance is the sum of a number of motions called modes. While it is not necessary for each mode to be stable, it is necessary to know for each mode the stability characteristics and response characteristics. A mode can be unstable providing its response characteristics are such that the pilot can easily control the airplane. On the other hand, even if a mode is stable, its response characteristics must be such that the airplane handles well (handling qualities). The design problem is to ensure that an aircraft has desirable stability and response characteristics thoughout the flight envelope and for all allowable cg positions. During this part of the design process, it may no longer be possible to modify the configuration, and automatic control solutions may have to be used.

App. A contains the geometric and aerodynamic data used in the text to compute performance, stability and control characteristics of a subsonic business jet called the SBJ throughout the text. App. B gives the relationship between the stability derivatives of Chap. 11 and the aerodynamics of Chap. 8. Finally, App. C contains a review of linear system theory for first-order systems and second-order systems.

1.6   Aircraft Sizing

While aircraft sizing is not covered in this text, it is useful to discuss the process to see where performance and static stability fit into the picture.

Consider the case of sizing a subsonic business jet to have a given range at a given cruise altitude. Furthermore, the aircraft must take-off and land on runways of given length and have a certain maximum rate of climb at the cruise altitude. The first step in the design process is to perform conceptual design. Here, the basic configuration is selected, which essentially means that a three-view drawing of the airplane can be sketched (no dimensions). The next step is to size the engines and the wing so that the mission can be performed. To size an engine, the performance of an actual engine is scaled up or down. See Fig. 1.8 for a flow chart of the sizing process. The end result of the sizing process is a three-view drawing of an airplane with dimensions.

The sizing process is iterative and begins by guessing the take-off gross weight, the engine size (maximum sea level static thrust), and the wing size (wing planform area). Next, the geometry of the airplane is determined by assuming that the center of gravity is located at the wing aerodynamic center, so that the airplane is statically stable. On the first iteration, statistical formulas are used to locate the horizontal and vertical tails. After the first iteration, component weights are available, and statistical formulas are used to place the tails. Once the geometry is known, the aerodynamics (drag polar) is estimated.

The next step is to fly the airplane through the mission. If the take-off distance is too large, the maximum thrust is increased, and the mission is restarted. Once take-off can be accomplished, the maximum rate of climb at the cruise altitude is determined. If it is less than the required value, the maximum thrust is increased, and the mission is restarted. The last constraint is landing distance. If the landing

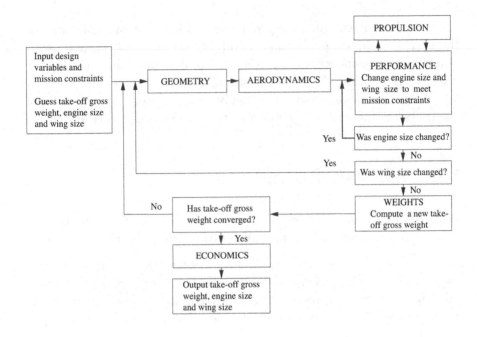

Figure 1.8: Aircraft Sizing Flowchart

distance is too large, the wing planform area is changed, and the mission is restarted. Here, however, the geometry and the aerodynamics must be recomputed.

Once the airplane can be flown through the entire mission, the amount of fuel required is known. Next, the fuel is allocated to wing, tip, and fuselage tanks, and statistical weights formulas are used to estimate the weight of each component and, hence, the take-off gross weight. If the computed take-off gross weight is not close enough to the guessed take-off gross weight, the entire process is repeated with the computed take-off gross weight as the guessed take-off gross weight. Once convergence has been achieved, the flyaway cost and the operating cost can be estimated.

1.7   Simulation

Simulations come in all sizes, but they are essentially computer programs that integrate the equations of motion. They are used to evaluate the flight characteristics of a vehicle. In addition to being run as computer programs, they can be used with working cockpits to allow pilots to evaluate handling qualities.

A major effort of an aerospace company is the creation of a high-fidelity 6DOF simulation for each of the vehicles it is developing. The simulation is modular in nature in that the aerodynamics function or subroutine is maintained by aerodynamicists, and so on.

Some performance problems, such as the spin, have so much interaction between the force and moment equations that they may have to be analyzed with six degree of freedom codes. These codes would essentially be simulations.

Chapter 2

3DOF Equations of Motion

An airplane operates near the surface of the earth which moves about the sun. Suppose that the equations of motion ($F = ma$ and $M = I\alpha$) are derived for an accurate inertial reference frame and that approximations characteristic of airplane flight (altitude and speed) are introduced into these equations. What results is a set of equations which can be obtained by assuming that the earth is flat, nonrotating, and an approximate inertial reference frame, that is, the flat earth model.

The equations of motion are composed of translational (force) equations ($F = ma$) and rotational (moment) equations ($M = I\alpha$) and are called the six degree of freedom (6DOF) equations of motion. For trajectory analysis (performance), the translational equations are uncoupled from the rotational equations by assuming that the airplane rotational rates are small and that control surface deflections do not affect forces. The translational equations are referred to as the three degree of freedom (3DOF) equations of motion.

As discussed in Chap. 1, two important legs of the commercial and military airplane missions are the climb and the cruise which occur in a vertical plane (a plane perpendicular to the surface of the earth). The purpose of this chapter is to derive the 3DOF equations of motion for flight in a vertical plane over a flat earth. First, the physical model is defined; several reference frames are defined; and the angular positions and rates of these frames relative to each other are determined. Then, the kinematic, dynamic, and weight equations are derived and discussed for nonsteady and quasi-steady flight. Next, the equations of motion for flight over a spherical earth are examined to find out how good the flat earth model really is. Finally, motivated by such problems as flight in a headwind, flight in the downwash of a tanker, and flight through a downburst, the equations of motion for flight in a moving atmosphere are derived.

2.1 Assumptions and Coordinate Systems

In deriving the equations of motion for the nonsteady flight of an airplane in a vertical plane over a flat earth, the following physical model is assumed:

a. The earth is flat, nonrotating, and an approximate inertial reference frame. The acceleration of gravity is constant and perpendicular to the surface of the earth. This is known as the *flat earth model*.

b. The atmosphere is at rest relative to the earth, and atmospheric properties are functions of altitude only.

c. The airplane is a conventional jet airplane with fixed engines, an aft tail, and a right-left *plane of symmetry*. It is modeled as a variable-mass particle.

d. The forces acting on an airplane in symmetric flight (no sideslip) are the thrust, the aerodynamic force, and the weight. They act at the center of gravity of the airplane, and the thrust and the aerodynamic force lie in the plane of symmetry.

The derivation of the equations of motion is clarified by defining a number of coordinate systems. For each coordinate system that moves with the airplane, the x and z axes are in the plane of symmetry of the airplane, and the y axis is such that the system is right handed. The x axis is in the direction of motion, while the z axis points earthward if the aircraft is in an upright orientation. Then, the y axis points out the right wing (relative to the pilot). The four coordinate systems used here are the following (see Fig. 2.1):

a. The *ground axes system* Exyz is fixed to the surface of the earth at mean sea level, and the xz plane is the vertical plane. It is an approximate inertial reference frame.

b. The *local horizon axes system* $Ox_h y_h z_h$ moves with the airplane (O is the airplane center of gravity), but its axes remain parallel to the ground axes.

c. The *wind axes system* $Ox_w y_w z_w$ moves with the airplane, and the x_w axis is coincident with the velocity vector.

d. The *body axes system* $Ox_b y_b z_b$ is fixed to the airplane.

These coordinate systems and their orientations are the convention in flight mechanics (see, for example, Ref. Mi1).

The coordinate systems for flight in a vertical plane are shown in Fig. 2.1, where the airplane is located at an altitude h above mean sea level. In the figure, **V** denotes the velocity of the airplane relative to the

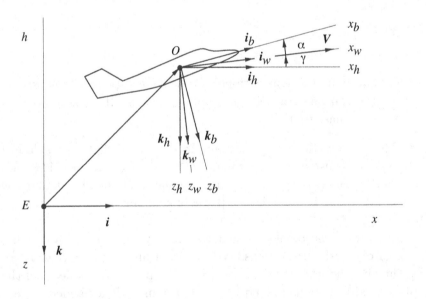

Figure 2.1: Coordinate Systems for Flight in a Vertical Plane

air; however, since the atmosphere is at rest relative to the ground, **V** is also the velocity of the airplane relative to the ground. Note that the wind axes are orientated relative to the local horizon axes by the *flight path angle* γ, and the body axes are orientated relative to the wind axes by the *angle of attack* α.

The unit vectors associated with the coordinate directions are denoted by $\mathbf{i}, \mathbf{j}$, and $\mathbf{k}$ with appropriate subscripts. Since the local horizon axes are always parallel to the ground axes, their unit vectors are equal, that is,

$$\mathbf{i}_h = \mathbf{i}$$
$$\mathbf{k}_h = \mathbf{k} \ . \tag{2.1}$$

Next, the wind axes unit vectors are related to the local horizon unit vectors as

$$\mathbf{i}_w = \cos\gamma\mathbf{i}_h - \sin\gamma\mathbf{k}_h$$
$$\mathbf{k}_w = \sin\gamma\mathbf{i}_h + \cos\gamma\mathbf{k}_h \ . \tag{2.2}$$

Since the unit vectors (2.1) are constant (fixed magnitude and direction), it is seen that

$$\frac{d\mathbf{i}_h}{dt} = \frac{d\mathbf{i}}{dt} = 0$$
$$\frac{d\mathbf{k}_h}{dt} = \frac{d\mathbf{k}}{dt} = 0 \ . \tag{2.3}$$

Then, by straight differentiation of Eqs. (2.2), the following relations are obtained:

$$\frac{d\mathbf{i}_w}{dt} = -\dot{\gamma}\mathbf{k}_w$$
$$\frac{d\mathbf{k}_w}{dt} = \dot{\gamma}\mathbf{i}_w \ . \tag{2.4}$$

The body axes are used primarily to define the angle of attack and will be discussed later.

2.2   Kinematic Equations

Kinematics is used to derive the differential equations for x and h which locate the airplane center of gravity relative to the origin of the ground axes system (*inertial position*). The definition of velocity relative to the ground (*inertial velocity*) is given by

$$\mathbf{V} = \frac{d\mathbf{EO}}{dt} \ , \tag{2.5}$$

where the derivative is taken holding the ground unit vectors constant. The velocity $\mathbf{V}$ and the position vector $\mathbf{EO}$ must be expressed in the same coordinate system to obtain the corresponding scalar equations. Here, the local horizon system is used where

$$\begin{aligned} \mathbf{V} &= V\mathbf{i}_w = V\cos\gamma\mathbf{i}_h - V\sin\gamma\mathbf{k}_h \\ \mathbf{EO} &= x\mathbf{i} - h\mathbf{k} = x\mathbf{i}_h - h\mathbf{k}_h \ . \end{aligned} \tag{2.6}$$

Since the unit vectors $\mathbf{i}_h$ and $\mathbf{k}_h$ are constant, Eq. (2.5) becomes

$$V\cos\gamma\mathbf{i}_h - V\sin\gamma\mathbf{k}_h = \dot{x}\mathbf{i}_h - \dot{h}\mathbf{k}_h \tag{2.7}$$

and leads to the following scalar equations:

$$\begin{aligned} \dot{x} &= V\cos\gamma \\ \dot{h} &= V\sin\gamma \ . \end{aligned} \tag{2.8}$$

These equations are the *kinematic equations* of motion for flight in a vertical plane.

2.3   Dynamic Equations

Dynamics is used to derive the differential equations for V and γ which define the velocity vector of the airplane center of gravity relative to the ground. Newton's second law states that

$$\mathbf{F} = m\mathbf{a} \tag{2.9}$$

where $\mathbf{F}$ is the resultant external force acting on the airplane, m is the mass of the airplane, and $\mathbf{a}$ is the *inertial acceleration* of the airplane. For the normal operating conditions of airplanes (altitude and speed), a reference frame fixed to the earth is an approximate inertial frame. Hence, $\mathbf{a}$ is approximated by the acceleration of the airplane relative to the ground.

The resultant external force acting on the airplane is given by

$$\mathbf{F} = \mathbf{T} + \mathbf{A} + \mathbf{W} \tag{2.10}$$

where $\mathbf{T}$ is the *thrust*, $\mathbf{A}$ is the *aerodynamic force*, and $\mathbf{W}$ is the *weight*. These concentrated forces are the result of having integrated the distributed forces over the airplane and having moved them to the center of gravity with appropriate moments. Recall that the moments are not needed because the force and moment equations have been uncoupled. By definition, the components of the aerodynamic force parallel and perpendicular to the velocity vector are called the *drag* and the *lift* so that

$$\mathbf{A} = \mathbf{D} + \mathbf{L} . \tag{2.11}$$

These forces are shown in Fig. 2.2 where the thrust vector is orientated relative to the velocity vector by the angle ε which is referred to as the *thrust angle of attack.*

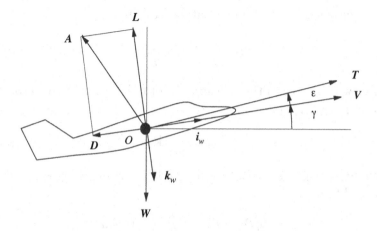

Figure 2.2: Forces Acting on an Airplane in Flight

In order to derive the scalar equations, it is necessary to select a coordinate system. While the local horizon system is used for obtaining the kinematic equations, a more direct derivation of the dynamic equations is possible by using the wind axes system. In this coordinate system, the forces acting on the airplane can be written as

$$
\begin{aligned}
\mathbf{T} &= T \cos\varepsilon \mathbf{i}_w - T \sin\varepsilon \mathbf{k}_w \\
\mathbf{D} &= -D \mathbf{i}_w \\
\mathbf{L} &= -L \mathbf{k}_w \\
\mathbf{W} &= -W \sin\gamma \mathbf{i}_w + W \cos\gamma \mathbf{k}_w
\end{aligned} \tag{2.12}
$$

so that the resultant external force becomes

$$\mathbf{F} = (T \cos\varepsilon - D - W \sin\gamma)\mathbf{i}_w - (T \sin\varepsilon + L - W \cos\gamma)\mathbf{k}_w . \tag{2.13}$$

By definition of acceleration relative to the ground,

$$\mathbf{a} = \frac{d\mathbf{V}}{dt} \tag{2.14}$$

holding the ground axes unit vectors constant. Since the velocity is along the x_w axis, it can be expressed as

$$\mathbf{V} = V\mathbf{i}_w \tag{2.15}$$

where both the velocity magnitude V and the direction of the unit vector $\mathbf{i}_w$ are functions of time. Differentiation leads to

$$\mathbf{a} = \dot{V}\mathbf{i}_w + V\frac{d\mathbf{i}_w}{dt}, \tag{2.16}$$

and in view of Eq. (2.4) the acceleration of the airplane with respect to the ground is given by

$$\mathbf{a} = \dot{V}\mathbf{i}_w - V\dot{\gamma}\mathbf{k}_w. \tag{2.17}$$

By combining Eqs. (2.9), (2.13), and (2.17), the following scalar equations are obtained:

$$
\begin{aligned}
\dot{V} &= (g/W)(T\cos\varepsilon - D - W\sin\gamma) \\
\dot{\gamma} &= (g/WV)(T\sin\varepsilon + L - W\cos\gamma)
\end{aligned}
\tag{2.18}
$$

where g is the constant acceleration of gravity and where the relation $W = mg$ has been used.

For conventional aircraft, the engines are fixed to the aircraft. This means that the angle between the thrust vector and the x_b axis is constant. Hence, from Fig. 2.3, it is seen that

$$\varepsilon = \alpha + \varepsilon_0 \tag{2.19}$$

where ε_0 is the value of ε when $\alpha = 0$ or the angle which the engine centerline makes with the x_b axis. Then, Eqs. (2.18) can be rewritten as

$$
\begin{aligned}
\dot{V} &= (g/W)[T\cos(\alpha + \varepsilon_0) - D - W\sin\gamma] \\
\dot{\gamma} &= (g/WV)[T\sin(\alpha + \varepsilon_0) + L - W\cos\gamma]
\end{aligned}
\tag{2.20}
$$

and are the *dynamic equations* for flight in a vertical plane.

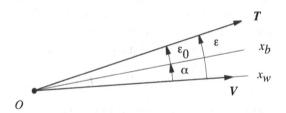

Figure 2.3: Relationship between ε and α

2.4   Weight Equation

By definition of the fuel weight flow rate $\dot{W}_{fuel}$, the rate of change of the weight of the aircraft is given by

$$\dot{W} = -\dot{W}_{fuel} \qquad (2.21)$$

Next, the *specific fuel consumption*

$$C = \frac{\dot{W}_{fuel}}{T} \qquad (2.22)$$

is introduced because it has some special properties. The *weight equation* becomes

$$\dot{W} = -CT. \qquad (2.23)$$

and gives the rate at which the weight of the aircraft is changing in terms of the operating conditions of the propulsion system.

2.5   Discussion of 3DOF Equations

The equations of motion for nonsteady flight in a vertical plane over a flat earth are given by Eqs. (2.8), (2.20), and (2.23), that is,

$$
\begin{aligned}
\dot{x} &= V\cos\gamma \\
\dot{h} &= V\sin\gamma \\
\dot{V} &= (g/W)[T\cos(\alpha + \varepsilon_0) - D - W\sin\gamma] \qquad (2.24)\\
\dot{\gamma} &= (g/WV)[T\sin(\alpha + \varepsilon_0) + L - W\cos\gamma] \\
\dot{W} &= -CT
\end{aligned}
$$

where g and ε_0 are constants. The purpose of this discussion is to examine the system of equations to see if it can be solved. For a fixed
geometry airplane in free flight with flaps up and gear up, it is known
that drag and lift obey functional relations of the form (see Chap. 3)

$$D = D(h, V, \alpha) , \quad L = L(h, V, \alpha). \tag{2.25}$$

It is also known that thrust and specific fuel consumption satisfy functional relations of the form (see Chap. 3)

$$T = T(h, V, P), \quad C = C(h, V, P) . \tag{2.26}$$

In these functional relations, V is the velocity of the airplane relative to
the atmosphere. However, since the atmosphere is fixed relative to the
earth, V is also the velocity of the airplane relative to the earth. The
quantity P is the engine *power setting*. As a consequence, the equations
of motion (2.24) involve the following variables:

$$x(t), h(t), V(t), \gamma(t), W(t), P(t), \alpha(t) . \tag{2.27}$$

The variables x, h, V, γ and W whose derivatives appear in the equations
of motion are called *state variables*. The remaining variables α and P
whose derivatives do not appear are called *control variables*.

Actually, the pilot controls the airplane by moving the throttle
and the control column. When the pilot moves the throttle, the fuel
flow rate to the engine is changed resulting in a change in the rpm of
the engine. The power setting of a jet engine is identified with the
ratio of the actual rpm to the maximum allowable rpm. Hence, while
the pilot actually controls the throttle angle, he can be thought of as
controlling the relative rpm of the engine. Similarly, when the pilot pulls
back on the control column, the elevator rotates upward, and the lift of
the horizontal tail increases in the downward sense. This increment of
lift creates an aerodynamic moment about the center of gravity which
rotates the airplane nose-up, thereby increasing the airplane angle of
attack. Hence, the pilot can be thought of as controlling the angle of
attack of the airplane rather than the angle of the control column. In
conclusion, for the purpose of computing the trajectory of an airplane,
the power setting and the angle of attack are identified as the controls.

The number of *mathematical degrees of freedom* of a system
of equations is the number of variables minus the number of equations.
The system of equations of motion (2.24) involves seven variables, five

equations, and two mathematical degrees of freedom. Hence, the time histories of two variables must be specified before the system can be integrated. This makes sense because there are two independent controls available to the pilot. On the other hand, it is not necessary to specify the control variables, as any two variables or any two relations between existing variables will do. For example, instead of flying at constant power setting and constant angle of attack, it might be necessary to fly at constant altitude and constant velocity. As another example, it might be desired to fly at constant power setting and constant dynamic pressure $\bar{q} = \rho(h)V^2/2$. In all, two additional equations involving existing variables must be added to complete the system.

In addition to the extra equations, it is necessary to provide a number of boundary conditions. Since the equations of motion are first-order differential equations, the integration leads to five constants of integration. One way to determine these constants is to specify the values of the state variables at the initial time, which is also prescribed. Then, to obtain a finite trajectory, it is necessary to give one final condition. This integration problem is referred to as an initial-value problem. If some variables are specified at the initial point and some variables are specified at the final point, the integration problem is called a boundary-value problem.

In conclusion, if the control action of the pilot or an equivalent set of relations is prescribed, the trajectory of the aircraft can be found by integrating the equations of motion subject to the boundary conditions. The trajectory is the set of functions $X(t)$, $h(t)$, $V(t)$, $\gamma(t)$, $W(t)$, $P(t)$ and $\alpha(t)$.

In airplane performance, it is often convenient to use lift as a variable rather than the angle of attack. Hence, if the expression for the lift (2.25) is solved for the angle of attack and if the angle of attack is eliminated from the expression for the drag (2.25), it is seen that

$$\alpha = \alpha(h, V, L), \quad D = D(h, V, L). \tag{2.28}$$

If these functional relations are used in the equations of motion (2.24), the lift becomes a control variable in place of the angle attack. It is also possible to write the engine functional relations in the form $P = P(h, V, T)$ and $C = C(h, V, T)$.

Because the system of Eqs. (2.24) has two mathematical degrees of freedom, it is necessary to provide two additional equations relating existing variables before the equations can be solved. With this model, the solution is usually numerical, say by using Runge-Kutta integration. Another approach is to use the degrees of freedom to optimize some performance capability of the airplane. An example is to minimize the time to climb from one altitude to another. The conditions to be satisfied by an optimal trajectory are derived in Ref. Hu. Optimization using this model is numerical in nature.

2.6   Quasi-Steady Flight

Strictly speaking, *quasi-steady flight* is defined by the approximations that the accelerations $\dot{V}$ and $\dot{\gamma}$ are negligible. However, for the performance problems to be analyzed (climb, cruise, descent) additional approximations also hold. They are small flight path inclination, small angle of attack and hence small thrust angle of attack, and small component of the thrust normal to the flight path. The four approximations which define quasi-steady flight are written as follows:

1. $\dot{V}, \dot{\gamma}$ negligible

2. $\gamma^2 << 1$ or $\cos\gamma \cong 1$, $\sin\gamma \cong \gamma$

3. $\varepsilon^2 << 1$ or $\cos\varepsilon \cong 1$, $\sin\varepsilon \cong \varepsilon$

4. $T\varepsilon << W$

For those segments of an airplane mission where the accelerations are negligible, the equations of motion become

$$\dot{x} = V$$
$$\dot{h} = V\gamma$$
$$0 = T - D - W\gamma \qquad (2.29)$$
$$0 = L - W$$
$$\dot{W} = -CT.$$

Note that if the drag is expressed as $D = D(h, V, L)$ the angle of attack no longer appears in the equations of motion. This means that only the drag is needed to represent the aerodynamics. Note also that the approximations do not change the number of mathematical degrees of freedom; there are still two. However, there are now three states (x, h, W) and four controls (V, γ, P, L).

Because two of the equations of motion are algebraic, they can be solved for two of the controls as

$$L = W \qquad (2.30)$$

and

$$\gamma = \frac{T(h, V, P) - D(h, V, W)}{W} . \qquad (2.31)$$

Then, the differential equations can be rewritten as

$$\begin{aligned}
\frac{dx}{dt} &= V \\
\frac{dh}{dt} &= V \left[\frac{T(h,V,P) - D(h,V,W)}{W} \right] \\
\frac{dW}{dt} &= -C(h, V, P) T(h, V, P)
\end{aligned} \qquad (2.32)$$

and still have two mathematical degrees of freedom (V,P).

The time is a good variable of integration for finding numerical solutions of the equations of motion. However, a goal of flight mechanics is to find analytical solutions. Here, the variable of integration depends on the problem. For climbing flight from one altitude to another, the altitude is chosen to be the variable of integration. The states then become x, t, W. To change the variable of integration in the equations of motion, the states are written as

$$x = x(h(t)) , \quad t = t(h(t)) , \quad W = W(h(t)) . \qquad (2.33)$$

Next, these expressions are differentiated with respect to the time as

$$\frac{dx}{dt} = \frac{dx}{dh}\frac{dh}{dt}, \quad \frac{dt}{dt} = \frac{dt}{dh}\frac{dh}{dt}, \quad \frac{dW}{dt} = \frac{dW}{dh}\frac{dh}{dt} \qquad (2.34)$$

and lead to

$$\frac{dx}{dh} = \frac{dx/dt}{dh/dt}, \quad \frac{dt}{dh} = \frac{1}{dh/dt}, \quad \frac{dW}{dh} = \frac{dW/dt}{dh/dt}. \qquad (2.35)$$

Note that the differentials which make up a derivative can be treated as algebraic quantities.

The equations of motion (2.32) with altitude as the variable of integration are given by

$$\frac{dx}{dh} = \frac{1}{\frac{T(h,V,P)-D(h,V,W)}{W}}$$

$$\frac{dt}{dh} = \frac{1}{V\left[\frac{T(h,V,P)-D(h,V,W)}{W}\right]} \qquad (2.36)$$

$$\frac{dW}{dh} = -\frac{C(h,V,P)T(h,V,P)}{V\left[\frac{T(h,V,P)-D(h,V,W)}{W}\right]}.$$

From this point on all the variables in these equations are considered to be functions of altitude. There are three states $x(h), t(h), W(h)$ and two controls $P(h), V(h)$ and still two mathematical degrees of freedom.

Eqs. (2.36) lead to optimization problems which can be handled quite easily. For example, suppose that it is desired to minimize the time to climb from one altitude to another. Because the amount of fuel consumed during the climb is around 5% of the initial climb weight, the weight on the right hand side of Eqs. (2.36) can be assumed constant. Also, an efficient way to climb is with maximum continuous power setting (constant power setting). The former assumption is an engineering approximation, while the latter is a statement as to how the airplane is being flown and reduces the number of degrees of freedom to one. With these assumptions, the second of Eqs. (2.36) can be integrated to obtain

$$t_f - t_0 = \int_{h_0}^{h_f} f(h,V)\ dh, \quad f(h,V) = \frac{1}{V\left[\frac{T(h,V,P)-D(h,V,W)}{W}\right]} \qquad (2.37)$$

because P and W are constant.

The problem of finding the function $V(h)$ which minimizes the time is a problem of the calculus of variations or optimal control theory (see Chap. 8 of Ref. Hu). However, because of the form of Eq. (2.37), it is possible to bypass optimization theory and get the optimal $V(h)$ by simple reasoning.

To find the velocity profile $V(h)$ which minimizes the time to climb, it is necessary to minimize the integral (2.37) with respect to the velocity profile $V(h)$. The integral is the area under a curve. The way to minimize the area under the curve is to minimize f with respect to V at each value of h. Hence, the conditions to be applied for finding the minimal velocity profile $V(h)$ are the following:

$$\left.\frac{\partial f}{\partial V}\right|_{h=Const} = 0, \quad \left.\frac{\partial^2 f}{\partial V^2}\right|_{h=Const} > 0. \qquad (2.38)$$

Substitution of the optimal velocity profile $V(h)$ into Eq. (2.37) leads to an integral of the form $\int g(h)dh$. If the aerodynamic and propulsion data are tabular, that is, interpolated tables of numbers, the optimization process must be carried out numerically. On the other hand, if an analytical model is available for the aerodynamics and propulsion, it may be possible to obtain the optimal velocity profile analytically.

Instead of optimizing the climb, it is possible to specify the velocity, say constant velocity. Then, the integration in Eq. (2.37) can be carried out for the time. A question which still remains is what constant velocity should be flown such that the time takes on a best value.

In finding the $V(h)$ which minimizes the time, all possible velocity profiles are in contention for the minimum. In finding the constant velocity which gives the best time, only constant velocity profiles are in contention. The velocity profile which minimizes the time has a lower value of the time than the best constant velocity climb.

2.7 Three-Dimensional Flight

In general, the velocity vector is oriented relative to the body axes by the sideslip angle and the angle of attack. If the velocity vector is in the plane of symmetry of the airplane, the sideslip angle is zero. Such flight is called symmetric. For three-dimensional, symmetric flight over a flat earth, the equations of motion are given by (Ref. Mi1)

$$
\begin{aligned}
\dot{x} &= V \cos\gamma \cos\psi \\
\dot{y} &= V \cos\gamma \sin\psi \\
\dot{h} &= V \sin\gamma \\
\dot{V} &= (g/W)[T\cos\varepsilon - D - W\sin\gamma] \\
\dot{\psi} &= (g/WV\cos\gamma)(T\sin\varepsilon + L)\sin\mu \\
\dot{\gamma} &= (g/WV)[(T\sin\varepsilon + L)\cos\mu - W\cos\gamma] \\
\dot{W} &= -CT .
\end{aligned}
\tag{2.39}
$$

In general (see Fig. 2.4), ψ is called *velocity yaw*; γ is called *velocity pitch*; and μ is called *velocity roll*. ψ is also called the *heading angle*, and

μ is called the *bank angle*. These angles are shown in Fig. 2.8 for flight in a horizontal plane.

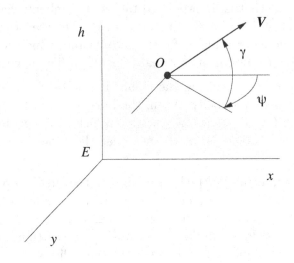

Figure 2.4: Three-Dimensional Flight

Note that if $\psi = 0$, these equations reduce to those for flight in a vertical plane (2.24), or if $\gamma = 0$, they reduce to those for flight in a horizontal plane (see Prob. 2.9).

Note also that the equations for $\dot{h}$ and $\dot{V}$ are the same as those for flight in a vertical plane. This result will be used in Chap. 7 when studying energy-maneuverability.

2.8   Flight over a Spherical Earth

The situation for flight in a vertical plane over a nonrotating spherical earth is shown in Fig. 2.9. Here, r_s is the radius of the surface of the earth, x is a curvilinear coordinate along the surface of the earth, and the angular velocity of the earth is sufficiently small that it can be neglected in the analysis of airplane trajectories. The equations of motion are given by (see Prob. 2.10)

$$\dot{x} \;=\; r_s V \cos\gamma/(r_s + h) \qquad\qquad (2.40)$$

$$\dot{h} \;=\; V \sin\gamma \qquad\qquad (2.41)$$

$$\dot{V} = (1/m)[T\cos\varepsilon - D - mg\sin\gamma] \tag{2.42}$$

$$\dot{\gamma} = (1/mV)[T\sin\varepsilon + L - mg\cos\gamma] + V\cos\gamma/(r_s + h) \tag{2.43}$$

$$\dot{m} = -CT/g . \tag{2.44}$$

For a spherical earth, the *inverse-square gravitational law* is

$$g = g_s \left(\frac{r_s}{r_s + h}\right)^2 \tag{2.45}$$

where g_s is the acceleration of gravity at sea level.

To determine the ranges of values of h and V for which the spherical equations reduce to the flat earth equations, the extra terms

$$\frac{r_s}{r_s + h}, \quad \left(\frac{r_s}{r_s + h}\right)^2, \quad \frac{V\cos\gamma}{r_s + h} \tag{2.46}$$

are examined. Use is made of the *binomial expansion* (Taylor series for small z):

$$(1 + z)^n \cong 1 + nz . \tag{2.47}$$

The second term is more restrictive than the first. It is rewritten as

$$\left(\frac{r_s}{r_s + h}\right)^2 = \left(1 + \frac{h}{r_s}\right)^{-2} \cong 1 - \frac{2h}{r_s} \tag{2.48}$$

and requires that

$$2\frac{h}{r_s} \ll 1 \tag{2.49}$$

to obtain the corresponding flat earth term. The third term is compared with the gravitational term $-(g\cos\gamma)/V$. The sum of the two terms is rewritten as

$$-\frac{g\cos\gamma}{V}\left(1 - \frac{V^2}{gr}\right) \tag{2.50}$$

and with the use of Eq. (2.49) requires that

$$\left(\frac{V}{\sqrt{g_s r_s}}\right)^2 \ll 1 \tag{2.51}$$

to reduce to the flat earth term. Hence, if the approximations

$$\frac{2h}{r_s} \ll 1 \tag{2.52}$$

and

$$\left(\frac{V}{\sqrt{g_s r_s}}\right)^2 << 1, \tag{2.53}$$

are introduced into the spherical earth equations, they reduce to the equations of motion for flight over a flat earth. The second approximation means that the vehicle speed is much less than the *satellite speed* at the surface of the earth.

For the remaining analysis, it is assumed that

$$r_s = 20,900,000 \text{ ft}, \quad g_s = 32.2 \text{ ft/s}^2. \tag{2.54}$$

If a mile has 5,280 ft, the radius of the earth is 3,960 mi. Also, the satellite speed is 25,900 ft/s.

If it is assumed that $0.01 << 1$, these inequalities imply that the upper altitude and velocity values for which the flat earth model is valid are 105,000 ft and 2,600 ft/s, respectively. Hence, the flat earth model is very accurate for almost all aircraft. On the other hand, if it is assumed that $0.1 << 1$, the limits become h=1,050,000 ft and V=8,320 ft/s. Note that the height of the atmosphere is only around h=400,000 ft. In this case the flat earth model is valid throughout the atmosphere up to a Mach number around $M = 8$.

In closing it is observed that the flat-earth model is really flight over a spherical earth at airplane altitudes and speeds.

2.9   Flight in a Moving Atmosphere

This section is motivated by such problems as flight in a headwind (tailwind), flight of a refueling airplane in the downwash of the tanker, or flight close to the ground (takeoff or landing) through a microburst (downburst) which is a downward and outward flow of air below a thunderstorm. In the horizontal plane which is not covered here, an example problem is flight through regions of high and low pressure where the atmosphere is rotating about the center of the region.

With reference to Fig. 2.5, let $\mathbf{V}_o$ denote the velocity of the airplane relative to the ground axes system (the inertial system). Then, the kinematic equations of motion are obtained from

$$\frac{d\mathbf{EO}}{dt} = \mathbf{V}_o \tag{2.55}$$

and the dynamic equations of motion, from

$$\mathbf{F} = m\mathbf{a}_o = m\frac{d\mathbf{V}_o}{dt}.$$

(2.56)

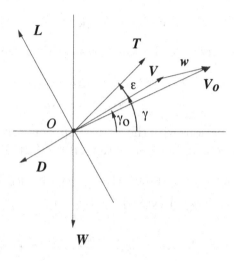

Figure 2.5: Flight in a Moving Atmosphere

Before deriving the scalar equations of motion, the velocity of the airplane relative to the ground ($\mathbf{V}_o$) is written as the vector sum of the velocity of the airplane relative to the atmosphere ($\mathbf{V}$) and the velocity of the atmosphere relative to the ground ($\mathbf{w}$), that is,

$$\mathbf{V}_o = \mathbf{V} + \mathbf{w}.$$

(2.57)

This is done because the drag, the lift, the thrust, and the specific fuel consumption are functions of the velocity of the airplane relative to the air. If the velocity of the atmosphere is written as

$$\mathbf{w} = w_x(x, h)\mathbf{i} - w_h(x, h)\mathbf{k},$$

(2.58)

the kinematic equations become

$$\dot{x} = V\cos\gamma + w_x$$

(2.59)

$$\dot{h} = V\sin\gamma + w_h.$$

(2.60)

Next, the acceleration can be written as

$$\mathbf{a}_o = \dot{\mathbf{V}}_o = \dot{\mathbf{V}} + \dot{\mathbf{w}}$$

(2.61)

In the wind axes frame, $\mathbf{V} = V\mathbf{i}_w$ (see Sec. 2.3) so that

$$\dot{\mathbf{V}} = \dot{V}\mathbf{i}_w - V\dot{\gamma}\mathbf{k}_w. \tag{2.62}$$

The acceleration of the atmosphere is obtained from Eq. (2.58) as

$$\dot{\mathbf{w}} = \dot{w}_x\mathbf{i} - \dot{w}_h\mathbf{k} \tag{2.63}$$

and can be transformed into the wind axes by using the relations

$$\begin{aligned}
\mathbf{i} &= \cos\gamma\,\mathbf{i}_w + \sin\gamma\,\mathbf{k}_w \\
\mathbf{k} &= -\sin\gamma\,\mathbf{i}_w + \cos\gamma\,\mathbf{k}_w.
\end{aligned} \tag{2.64}$$

Combining these equations with the force components given in Eq. (2.13) leads to the dynamic equations of motion (Ref. Mi2)

$$\begin{aligned}
\dot{V} &= \tfrac{g}{W}(T\cos\varepsilon - D - W\sin\gamma) - (\dot{w}_x\cos\gamma + \dot{w}_h\sin\gamma) \\
\dot{\gamma} &= \tfrac{g}{WV}(T\sin\varepsilon + L - W\cos\gamma) + \tfrac{1}{V}(\dot{w}_x\sin\gamma - \dot{w}_h\cos\gamma)
\end{aligned} \tag{2.65}$$

where

$$\begin{aligned}
\dot{w}_x &= \tfrac{\partial w_x}{\partial x}(V\cos\gamma + w_x) + \tfrac{\partial w_x}{\partial h}(V\sin\gamma + w_h) \\
\dot{w}_h &= \tfrac{\partial w_h}{\partial x}(V\cos\gamma + w_x) + \tfrac{\partial w_h}{\partial h}(V\sin\gamma + w_h)
\end{aligned} \tag{2.66}$$

As an example of applying these equations, consider flight into a constant headwind. Here, if V_w denotes the velocity of the headwind,

$$w_x = -V_w = \text{Const}, \quad \mathrm{w_h} = 0. \tag{2.67}$$

so that the kinematic equations are given by

$$\begin{aligned}
\dot{x} &= V\cos\gamma - V_w \\
\dot{h} &= V\sin\gamma
\end{aligned} \tag{2.68}$$

Note that the headwind only affects the horizontal distance as it should. Finally, the dynamic equations reduce to Eqs. (2.18).

As another example, consider the flight of a refueling airplane in the downwash behind the tanker. Here, the wind velocity is given by

$$w_x = 0, \quad w_h = -V_T\varepsilon = \text{Const} \tag{2.69}$$

where V_T is the velocity of the tanker and ε is the downwash angle (in the neighborhood of a couple of degrees). The kinematic equations become

$$\begin{aligned}
\dot{x} &= V\cos\gamma \\
\dot{h} &= V\sin\gamma - V_T\varepsilon.
\end{aligned} \tag{2.70}$$

In order for the refueling airplane to maintain constant altitude relative to the ground ($\dot{h} = 0$) during refueling, it must establish a small rate of climb relative to the atmosphere.

In Ref. Mi2, the microburst has been modeled by $w_x(x)$ and $w_h(x)$ as shown in Fig. 2.6. Here, w_x transitions from a headwind ($w_x = -40$ ft/s) to a tailwind ($w_x = 40$ ft/s) over a finite distance. The downwind is constant ($w_h = -20$ ft/s) over the central part of the downburst. Because there is a change in wind speed with position, there is windshear. If the pilot does not react quickly to the downburst, the airplane could lose altitude after the switch from the headwind to the tailwind and crash.

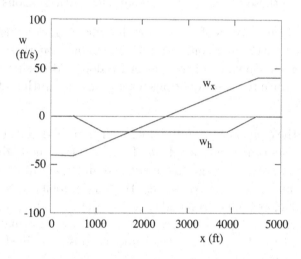

Figure 2.6: Approximate Wind Profiles in a Downburst

Problems

The first five problems are designed to give the student practice in deriving equations of motion in the wind axes system. Derive the equations of motion for the particular problem stated using the procedures of Sections 2.2, 2.3, and 2.4.

2.1 Derive the equations of motion for constant altitude flight. First, draw a free body diagram showing an aircraft in constant altitude flight and show all the coordinate systems, angles, and forces.

Show that these equations have one mathematical degree of freedom and that Eqs. (2.24) reduce to these equations when combined
with $h = $ Const.

2.2 Derive the equations of motion for flight at constant altitude and
constant velocity. Show that these equations have zero mathematical degrees of freedom. Also, show that Eqs. (2.24) reduce to
these equations when combined with $h=$Const and $V = $ Const.

2.3 Derive the equations of motion for nonsteady climbing flight at
constant flight path inclination. Show that these equations have
one mathematical degree of freedom. Also, show that Eqs. (2.24)
reduce to these equations when combined with $\gamma=$Const.

2.4 Derive the equations of motion for climbing flight at constant flight
path inclination and constant velocity. Show that these equations
have zero mathematical degrees of freedom. Also, show that Eqs.
(2.24) reduce to these equations when combined with $\gamma=$Const and
$V = $ Const.

2.5 Derive the equations of motion for an airplane in descending gliding
flight (T=0) in a vertical plane. First, draw a free body diagram
showing an aircraft in gliding flight and all the coordinate systems,
angles, and forces. Here, assume that the velocity vector is at an
angle ϕ below the horizon and that the aircraft is at a positive angle
of attack α. Show that these equations have one mathematical
degree of freedom and are the same as those obtained from Eqs.
(2.24) with $T = 0$ and $\gamma = -\phi$.

2.6 Derive the equations of motion in the local horizon axes system.
In other words, write the velocity vector as

$$\mathbf{V} = u\mathbf{i}_h - w\mathbf{k}_h,$$

and note that

$$V = \sqrt{u^2 + w^2}, \quad \tan\gamma = w/u \ .$$

Show that

$$\dot{x} = u$$
$$\dot{h} = w$$

$$\dot{u} = \frac{g}{W}\left[(T\cos\varepsilon - D)\frac{u}{\sqrt{u^2 + w^2}} - (T\sin\varepsilon + L)\frac{w}{\sqrt{u^2 + w^2}}\right]$$

$$\dot{w} = \frac{g}{W}\left[(T\cos\varepsilon - D)\frac{w}{\sqrt{u^2 + w^2}} + (T\sin\varepsilon + L)\frac{u}{\sqrt{u^2 + w^2}} - W\right]$$

$$\dot{W} = -CT$$

where $\varepsilon = \varepsilon_0 + \alpha$.

a. List the variables and show that there are two mathematical degrees of freedom.

b. Write

$$u = V\cos\gamma, \quad w = V\sin\gamma,$$

and show that the above equations can be manipulated into the equations of motion in the wind axes system (2.24).

2.7 Derive the equations of motion in the body axes system (see Fig. 2.7) where U, W are the components of the velocity vector on the body axes, Θ is the angle between the x_b axis and the x_h axis, called the pitch angle. Note that mg is used for the weight because W now denotes a velocity component.

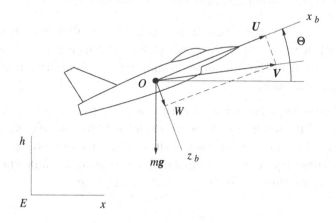

Figure 2.7: Equations of Motion in Body Axes

Write the velocity vector as

$$\mathbf{V} = U\mathbf{i}_b + W\mathbf{k}_b,$$

and show that

$$\dot{x} = U \cos \Theta + W \sin \Theta$$

$$\dot{h} = U \sin \Theta - W \cos \Theta$$

$$\dot{U} = -WQ + (1/m)[T \cos \varepsilon_0 + L \sin \alpha - D \cos \alpha - mg \sin \Theta]$$

$$\dot{W} = UQ - (1/m)[T \sin \varepsilon_0 + L \cos \alpha + D \sin \alpha - mg \cos \Theta]$$

$$\dot{\Theta} = Q$$

$$\dot{m}g = -CT.$$

where Q is the pitch rate. The velocity and the angle of attack satisfy the relations

$$V = \sqrt{U^2 + W^2}, \quad \tan \alpha = \frac{W}{U}.$$

List the variables, and show that there are two mathematical degrees of freedom.

2.8 Show that Eqs. (2.35) for three-dimensional flight over a flat earth have three mathematical degrees of freedom.

2.9 Consider the constant altitude turning flight (flight in a horizontal plane) of an airplane as shown in Fig. 2.8 where x, y denote the cg location in the horizontal plane, ψ is the heading angle, and μ is the bank angle.

a. Assuming a coordinated turn (no sideslip) so that the velocity, thrust, lift and drag vectors are in the airplane plane of symmetry, derive the equations of motion in the ground axes system. Show that these equations can be combined to form the following equations of motion for flight in a horizontal plane:

$$\dot{x} = V \cos \psi$$

$$\dot{y} = V \sin \psi$$

$$\dot{V} = (g/W)[T \cos(\varepsilon_0 + \alpha) - D]$$

$$\dot{\psi} = (g/WV)[T \sin(\varepsilon_0 + \alpha) + L] \sin \mu$$

$$0 = [T \sin(\varepsilon_0 + \alpha) + L] \cos \mu - W$$

$$\dot{W} = -CT.$$

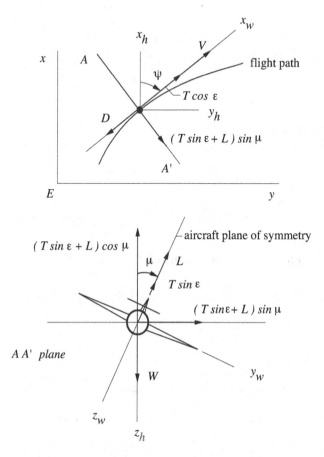

Figure 2.8: Nomenclature for Turning Flight

b. List the variables and show that these equations have two mathematical degrees of freedom.

2.10 The purpose of this exercise is to develop the equations of motion for flight in a great circle plane over a nonrotating spherical earth (see Fig. 2.9). Actually, the angular velocity of the earth is sufficiently small that it can be neglected in the analysis of vehicle performance. With reference to the figure, the earth axes system Exz is fixed to the surface of the earth. The z axis points toward the center of the earth, and the x axis is curvilinear $(x = r_s \lambda)$ and along the surface of the earth. Derive the equations of motion,

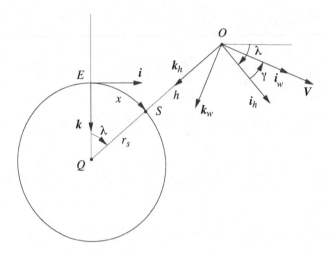

Figure 2.9: Nomenclature for Spherical Earth

that is,

$$\dot{x} = r_s V \cos\gamma / (r_s + h)$$

$$\dot{h} = V \sin\gamma$$

$$\dot{V} = (1/m)[T \cos\varepsilon - D - mg \sin\gamma]$$

$$\dot{\gamma} = (1/mV)[T \sin\varepsilon + L - mg \cos\gamma] + V \cos\gamma / (r_s + h)$$

$$\dot{m} = -CT/g$$

where, for a spherical earth, the inverse-square gravitational law is

$$g = g_s \left(\frac{r_s}{r_s + h} \right)^2 .$$

Note that $\mathbf{EO} = \mathbf{EQ} + \mathbf{QO}$ and that $\mathbf{EQ} = \text{const.}$

2.11 Derive the equations of motion in a moving atmosphere in the body axes system (review Sec. 2.9 and Prob. 2.7). Show that the equations of motion are

$$\dot{x} = U \cos\Theta + W \sin\Theta + w_x$$

$$\dot{h} = U \sin\Theta - W \cos\Theta + w_h$$

$$\dot{U} = -WQ + (1/m)[T \cos\varepsilon_0 + L \sin\alpha - D \cos\alpha - mg \sin\Theta]$$

$$-(\dot{w}_x \cos\Theta + \dot{w}_h \sin\Theta)$$
$$\dot{W} = UQ - (1/m)[T\sin\varepsilon_0 + L\cos\alpha + D\sin\alpha - mg\cos\Theta]$$
$$-(\dot{w}_x \sin\Theta - \dot{w}_h \cos\Theta)$$
$$\dot{\Theta} = Q$$
$$\dot{m}g = -CT.$$

where

$$V = \sqrt{U^2 + W^2}, \quad \tan\alpha = \frac{W}{U}.$$

Note that U, W are now the components of the velocity vector relative to the wind.

2.12 A sounding rocket is ascending vertically at zero angle of attack in the atmosphere over a flat earth (Fig. 2.10). Derive the kinematic

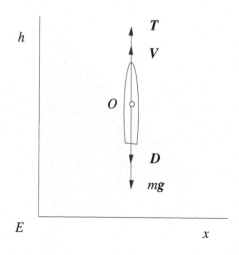

Figure 2.10: Sounding Rocket

and dynamic equations of motion. The thrust can be approximated by the relation $T = \beta c$ where β is the propellant mass flow rate (power setting) and c is the equivalent exhaust velocity (assumed constant). The mass equation is given by $\dot{m} = -\beta$. If $C_D = C_D(M)$, how many mathematical degrees of freedom do the equations of motion have? List the equations, the functional relations, and the variables.

Answer:

$$\dot{h} = V$$
$$\dot{V} = (1/m)[\beta c - D(h, V) - mg]$$
$$\dot{m} = -\beta$$

The variables are $h(t), V(t), m(t), \beta(t)$, so there is one mathematical degree of freedom. If β is constant (constant thrust), there are zero mathematical degrees of freedom.

Chapter 3

Atmosphere, Aerodynamics, and Propulsion

In the previous chapter, functional relations were presented for the aerodynamic and propulsion terms appearing in the equations of motion used for trajectory analysis. It is the intention of this chapter to verify these relations as well as to present a procedure for estimating the aerodynamic characteristics of a subsonic jet airplane. Engine data is assumed to be provided by the manufacturer, so that no estimation of propulsion terms is attempted.

First, a standard atmosphere is defined and approximated by an exponential atmosphere. Then, aerodynamics is discussed functionally, and an algorithm for estimating the angle of attack and the drag polar of a subsonic airplane at moderate values of the lift coefficient is presented. Because the graphs of these quantities have simple geometric forms, approximate aerodynamic formulas are developed. All of the aerodynamics figures are for the SBJ of App. A. Finally, data for a subsonic turbojet and turbofan are presented, and the propulsion terms are discussed functionally. Because of the behavior of these engines, approximate formulas can be developed.

3.1 Standard Atmosphere

The real atmosphere is in motion with respect to the earth, and its properties are a function of position (longitude, latitude, and altitude) and time. From an operational point of view, it is necessary to have this information, at least in the region of operation. However, from a design point of view, that is, when comparing the performance of two aircraft, it is only necessary that the atmospheric conditions be characteristic of the real atmosphere and be the same for the two airplanes. Hence, it is not important to consider the motion of the atmosphere or to vary its characteristics with respect to longitude and latitude. A simple model in which atmospheric properties vary with altitude is sufficient.

There are two basic equations which must be satisfied by air at rest: the aerostatic equation

$$dp = -\rho g \, dh \tag{3.1}$$

and the equation of state for a perfect gas

$$p = \rho R \tau \tag{3.2}$$

where p is the pressure, ρ the density, R the gas constant for air, and τ the absolute temperature. For the region of the atmosphere where airplanes normally operate, the acceleration of gravity and the composition of air can be assumed constant (g = 32.174 ft/s^2 and R = 1716.5 ft^2/s^2 °R). To complete the system of equations defining the *standard atmosphere*, it is assumed that the temperature is a known function of the altitude.

Actual measurements of atmospheric properties using balloons and sounding rockets have shown that the atmosphere can be approximated by a number of layers in which the temperature varies linearly with the altitude, that is, the temperature gradient $\beta = d\tau/dh$ is constant. The assumed temperature profile for the first three layers of the 1962 U.S. Standard Atmosphere (Ref. An) is shown in Fig. 3.1. Note that the layer of the atmosphere closest to the earth ($0 \le h \le 36089$ ft) is called the troposphere; the next two layers ($36089 \le h \le 104{,}990$ ft) are part of the stratosphere; and the dividing line between the troposphere and the stratosphere is called the tropopause.

Because of the assumed temperature profile, the equations defining *temperature*, *pressure*, and *density* can be written as

$$
\begin{aligned}
d\tau &= \beta \, dh \\
dp/p &= -(g/R)dh/\tau \\
d\rho/\rho &= -(g/R + \beta)dh/\tau
\end{aligned}
\tag{3.3}
$$

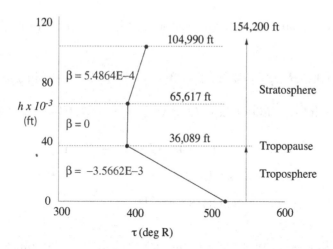

Figure 3.1: Temperature Distribution - 1962 U.S. Standard Atmosphere

where β is a constant for each layer of the atmosphere. For the *troposphere* ($\beta = -3.5662\text{E}{-}3$ °R/ft), these equations can be integrated to obtain

$$\tau = 518.69 - 3.5662\text{E}{-}3\,h$$

$$p = 1.1376\text{E}{-}11\,\tau^{5.2560} \qquad (3.4)$$

$$\rho = 6.6277\text{E}{-}15\,\tau^{4.2560}$$

where the *standard sea level conditions*

$$T_s = 518.69\,°R, \quad p_s = 2116.2\ \text{lb/ft}^2, \quad \rho_s = 2.3769\text{E}{-}3\ \text{slugs/ft}^3 \quad (3.5)$$

have been used to evaluate the constants of integration. The initial conditions for the first layer of the *stratosphere* are obtained by applying Eqs. (3.4) at the *tropopause* ($h = 36{,}089$ ft) and are given by

$$T_t = 389.99\,°R, \quad p_t = 472.68\ \text{lb/ft}^2, \quad \rho_t = 7.0613\text{E}{-}4\ \text{slugs/ft}^3. \quad (3.6)$$

Then, the integration of Eqs. (3.3) with $\beta = 0$ °R/ft leads to

$$\tau = 389.99$$

$$p = 2678.4\,\exp(-4.8063\text{E}{-}5\,h) \qquad (3.7)$$

$$\rho = 1.4939\text{E}{-}6\,p$$

Finally, the initial conditions for the second layer of the stratosphere are given by

$$\tau_+ = 389.99\ ^\circ R, \quad p_+ = 114.35\ \text{lb/ft}^2, \quad \rho_+ = 1.7083\text{E--}4\ \text{slugs/ft}^3 \quad (3.8)$$

so that the integration of Eqs. (3.3) with $\beta = 5.4864\text{E--}4\ ^\circ R/\text{ft}$ yields

$$
\begin{aligned}
\tau &= 389.99 + 5.4864\text{E--}4\ (h - 65,617)\ , \\
p &= 3.7930\text{E+}90\ \tau^{-34.164} \\
\rho &= 2.2099\text{E+}87\ \tau^{-35.164}
\end{aligned}
\quad (3.9)
$$

Two other properties of interest are the speed of sound and the viscosity. If the ratio of specific heats for air is denoted by k ($k = 1.4$), the *speed of sound* is given by

$$a = (kR\tau)^{1/2} = 49.021\tau^{1/2}. \quad (3.10)$$

The *viscosity* is assumed to satisfy Sutherland's formula

$$\mu = \frac{2.27\text{E--}8\ \tau^{3/2}}{\tau + 198.6}. \quad (3.11)$$

Since the absolute temperature is a known function of altitude, speed of sound and viscosity become functions of altitude.

The end result of this analysis is that the atmospheric properties satisfy functional relations of the form $\tau = \tau(h)$, $p = p(h)$, $\rho = \rho(h)$, $a = a(h)$ and $\mu = \mu(h)$. Values of these quantities are presented in Table 3.1 for the standard day.

The same quantities can be derived for a nonstandard day by shifting the temperature profile in Fig. 3.1 to the right or to the left by the amount $\Delta\tau$ Then, the equations for the atmosphere must be rederived.

3.2   Exponential Atmosphere

Outside the first layer of the stratosphere, the formulas for the atmospheric properties given by the standard atmosphere are so complicated mathematically (decimal exponents) that their use will not lead to analytical solutions. An approximate atmosphere which may lead to analytical solutions is the *exponential atmosphere* or *isothermal atmosphere*. Here, the formula for the density is given by

$$\rho = \rho_s \exp(-h/\lambda) \qquad (3.12)$$

where ρ_s is the sea level density and λ is called the scale height. This form is motivated by the statosphere formulas where the temperature is constant and exponential is exact. For the troposphere and the constant temperature part of the stratosphere, a value of λ which gives an error on the order of 10% is $\lambda = 26,600$ ft.

To achieve more accuracy, it is possible to assume that each layer of the atmosphere satisfies an exponential form. Here,

Table 3.1: Atmospheric Properties versus Altitude

h ft	τ °R	p lb/ft^2	ρ slug/ft^3	a ft/s	μ lb s/ft^2
0.	518.69	2116.1	2.3769E-03	1116.4	3.7385E-07
1000.	515.12	2040.8	2.3081E-03	1112.6	3.7185E-07
2000.	511.56	1967.6	2.2409E-03	1108.7	3.6984E-07
3000.	507.99	1896.6	2.1751E-03	1104.9	3.6783E-07
4000.	504.43	1827.6	2.1109E-03	1101.0	3.6581E-07
5000.	500.86	1760.7	2.0481E-03	1097.1	3.6378E-07
6000.	497.29	1695.8	1.9868E-03	1093.2	3.6174E-07
7000.	493.73	1632.9	1.9268E-03	1089.2	3.5970E-07
8000.	490.16	1571.8	1.8683E-03	1085.3	3.5766E-07
9000.	486.59	1512.7	1.8111E-03	1081.3	3.5560E-07
10000.	483.03	1455.3	1.7553E-03	1077.4	3.5354E-07
11000.	479.46	1399.7	1.7008E-03	1073.4	3.5147E-07
12000.	475.90	1345.8	1.6476E-03	1069.4	3.4939E-07
13000.	472.33	1293.7	1.5957E-03	1065.4	3.4731E-07
14000.	468.76	1243.1	1.5450E-03	1061.4	3.4522E-07
15000.	465.20	1194.2	1.4956E-03	1057.3	3.4312E-07
16000.	461.63	1146.9	1.4474E-03	1053.2	3.4101E-07
17000.	458.06	1101.1	1.4004E-03	1049.2	3.3890E-07
18000.	454.50	1056.8	1.3546E-03	1045.1	3.3678E-07
19000.	450.93	1013.9	1.3100E-03	1041.0	3.3465E-07
20000.	447.37	972.5	1.2664E-03	1036.8	3.3251E-07

Table 3.1: Atmospheric Properties versus Altitude (cont'd)

h	τ	p	ρ	a	μ
ft	°R	lb/ft^2	slug/ft^3	ft/s	lb s/ft^2
21000.	443.80	932.4	1.2240E-03	1032.7	3.3037E-07
22000.	440.23	893.7	1.1827E-03	1028.5	3.2822E-07
23000.	436.67	856.3	1.1425E-03	1024.4	3.2606E-07
24000.	433.10	820.2	1.1033E-03	1020.2	3.2389E-07
25000.	429.53	785.3	1.0651E-03	1016.0	3.2171E-07
26000.	425.97	751.6	1.0280E-03	1011.7	3.1953E-07
27000.	422.40	719.1	9.9187E-04	1007.5	3.1734E-07
28000.	418.84	687.8	9.5672E-04	1003.2	3.1514E-07
29000.	415.27	657.6	9.2253E-04	999.0	3.1293E-07
30000.	411.70	628.4	8.8928E-04	994.7	3.1071E-07
31000.	408.14	600.6	8.5695E-04	990.3	3.0849E-07
32000.	404.57	573.3	8.2553E-04	986.0	3.0625E-07
33000.	401.01	547.2	7.9501E-04	981.7	3.0401E-07
34000.	397.44	522.1	7.6535E-04	977.3	3.0176E-07
35000.	393.87	497.9	7.3654E-04	972.9	2.9950E-07
36000.	390.31	474.7	7.0858E-04	968.5	2.9723E-07
37000.	389.99	452.4	6.7589E-04	968.1	2.9703E-07
38000.	389.99	431.2	6.4418E-04	968.1	2.9703E-07
39000.	389.99	411.0	6.1395E-04	968.1	2.9703E-07
40000.	389.99	391.7	5.8514E-04	968.1	2.9703E-07
41000.	389.99	373.3	5.5768E-04	968.1	2.9703E-07
42000.	389.89	355.8	5.3151E-04	968.1	2.9703E-07
43000.	389.99	339.1	5.0657E-04	968.1	2.9703E-07
44000.	389.99	323.2	4.8280E-04	968.1	2.9703E-07
45000.	389.99	308.0	4.6014E-04	968.1	2.9703E-07
46000.	389.99	293.6	4.3855E-04	968.1	2.9703E-07
47000.	389.99	279.8	4.1797E-04	968.1	2.9703E-07
48000.	389.99	266.7	3.9835E-04	968.1	2.9703E-07
49000.	389.99	254.1	3.7966E-04	968.1	2.9703E-07
50000.	389.99	242.2	3.6184E-04	968.1	2.9703E-07
51000.	389.99	230.8	3.4486E-04	968.1	2.9703E-07
52000.	389.99	220.0	3.2868E-04	968.1	2.9703E-07
53000.	389.99	209.7	3.1326E-04	968.1	2.9703E-07
54000.	389.99	199.9	2.9856E-04	968.1	2.9703E-07
55000.	389.99	190.5	2.8455E-04	968.1	2.9703E-07
56000.	389.99	181.5	2.7119E-04	968.1	2.9703E-07
57000.	389.99	173.0	2.5847E-04	968.1	2.9703E-07
58000.	389.99	164.9	2.4634E-04	968.1	2.9703E-07
59000.	389.99	157.2	2.3478E-04	968.1	2.9703E-07
60000.	389.99	149.8	2.2376E-04	968.1	2.9703E-07

Table 3.1: Atmospheric Properties versus Altitude (cont'd)

h	τ	p	ρ	a	μ
ft	°R	lb/ft^2	slug/ft^3	ft/s	lb s/ft^2
61000.	389.99	142.8	2.1326E-04	968.1	2.9703E-07
62000.	389.99	136.1	2.0325E-04	968.1	2.9703E-07
63000.	389.99	129.7	1.9372E-04	968.1	2.9703E-07
64000.	389.99	123.6	1.8463E-04	968.1	2.9703E-07
65000.	389.99	117.8	1.7596E-04	968.1	2.9703E-07
66000.	390.20	112.3	1.6763E-04	968.3	2.9716E-07
67000.	390.75	107.0	1.5955E-04	969.0	2.9751E-07
68000.	391.30	102.0	1.5186E-04	969.7	2.9786E-07
69000.	391.85	97.2	1.4456E-04	970.4	2.9821E-07
70000.	392.39	92.7	1.3762E-04	971.1	2.9856E-07
71000.	392.94	88.4	1.3103E-04	971.7	2.9891E-07
72000.	393.49	84.3	1.2475E-04	972.4	2.9925E-07
73000.	394.04	80.3	1.1879E-04	973.1	2.9960E-07
74000.	394.59	76.6	1.1312E-04	973.8	2.9995E-07
75000.	395.14	73.1	1.0772E-04	974.4	3.0030E-07
76000.	395.69	69.7	1.0259E-04	975.1	3.0065E-07
77000.	396.24	66.5	9.7713E-05	975.8	3.0099E-07
78000.	396.78	63.4	9.3073E-05	976.5	3.0134E-07
79000.	397.33	60.5	8.8658E-05	977.1	3.0169E-07
80000.	397.88	57.7	8.4459E-05	977.8	3.0204E-07

$$
\begin{aligned}
\text{Troposphere} \quad & \rho = \rho_s \exp[-h/29,730] \\
\text{Stratosphere I} \quad & \rho = \rho_t \exp[-(h - h_t)/20,806] \qquad (3.13)\\
\text{Stratosphere II} \quad & \rho = \rho_+ \exp[-(h - h_+)/20,770]
\end{aligned}
$$

Note that the density for the first stratosphere layer is exact.

In the spirit of an isothermal atmosphere, the speed of sound can be assumed constant and have the value $a = 1000$ ft/s. This value is easy to remember and makes the conversion from Mach number to ft/s and vice versa simple.

3.3 Aerodynamics: Functional Relations

The resultant aerodynamic force is the integrated effect of the pressure and skin friction caused by the flow of air over the surface of the airplane.

The lift and the drag are the components of the resultant aerodynamic force perpendicular and parallel to the velocity vector. They satisfy the relations

$$L = \frac{1}{2}C_L\rho SV^2, \quad D = \frac{1}{2}C_D\rho SV^2 \tag{3.14}$$

where C_L is the *lift coefficient*, C_D is the *drag* coefficient, ρ is the density of the atmosphere at the altitude of the airplane, V is the velocity of the airplane relative to the atmosphere, and S is the *wing planform area*.

If the equations governing the motion of air (the continuity equation, the linear momentum equations, the energy equation, and the perfect gas equation) and the boundary conditions are nondimensional-ized, the integration of the pressure and skin friction coefficients over the surface of the airplane leads to the following functional relations for the lift coefficient and the drag coefficient for a constant geometry aircraft:

$$C_L = C_L(\alpha, M, R_e), \quad C_D = C_D(\alpha, M, R_e). \tag{3.15}$$

In these relations, α is the airplane angle of attack, while the *Mach number* and the *Reynolds number* are defined as

$$M = \frac{V}{a}, \quad R_e = \frac{\rho V l}{\mu}. \tag{3.16}$$

Here, a and μ are the speed of sound and the viscosity of the atmo-sphere at the altitude of the airplane, and l is a characteristic length of the airplane. In practice, Reynolds number effects are neglected in the expression for the lift coefficient so that Eqs. (3.15) can be rewritten as

$$C_L = C_L(\alpha, M), \quad C_D = C_D(\alpha, M, R_e). \tag{3.17}$$

In the calculation of aircraft performance, it is the convention to use lift coefficient as a variable rather than angle of attack. Hence, if the C_L equation is solved for α and the result is substituted into the expression for C_D , the following relations are obtained:

$$\alpha = \alpha(C_L, M), \quad C_D = C_D(C_L, M, R_e). \tag{3.18}$$

The equation for C_D is referred to as the *drag polar*.

Plots of the angle of attack and drag coefficient are shown in Fig. 3.2 for the Subsonic Business Jet (SBJ) in App. A. It is interesting to note that the lift coefficient has a maximum value. Also, the angle of

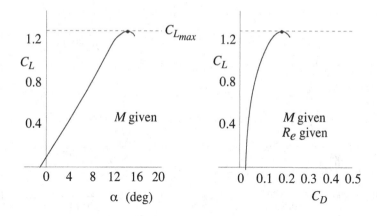

Figure 3.2: Angle of Attack and Drag Coefficient (SBJ)

attack is linear in the lift coefficient over a wide range of values of C_L, and the drag coefficient is parabolic over the same interval.

Dimensional expressions for the angle of attack and the drag can be obtained by combining Eqs. (3.14), (3.16), and (3.18) as

$$\alpha = \alpha[2L/\rho(h)SV^2, \; V/a(h)]$$

$$2D/\rho(h)SV^2 = C_D[2L/\rho(h)SV^2, \; V/a(h), \; \rho(h)Vl/\mu(h)]$$

$$(3.19)$$

so that

$$\alpha = \alpha(h, V, L), \quad D = D(h, V, L). \tag{3.20}$$

These are the expressions for the angle of attack and the drag used in Chap. 2 to discuss the solution of the quasi-steady equations of motion.

Plots of the angle of attack and the drag for the SBJ are shown in Fig. 3.3 for a given lift and several values of the altitude. Note that at each altitude the angle of attack decreases monotonically with the velocity and there is a velocity for minimum drag which increases with altitude.

Another important aerodynamic characteristic of an airplane is the *lift-to-drag ratio* or *aerodynamic efficiency*

$$E = \frac{L}{D} = \frac{C_L}{C_D}. \tag{3.21}$$

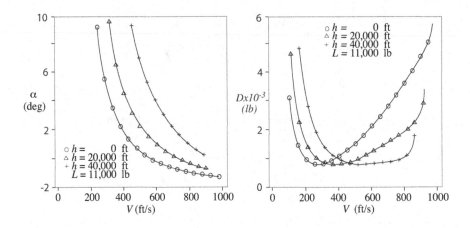

Figure 3.3: Angle of Attack and Drag for the SBJ

In terms of nondimensional variables, the lift-to-drag ratio satisfies the functional relation

$$E = E(C_L, M, R_e) \tag{3.22}$$

whereas the dimensional functional relation is given by

$$E = E(h, V, L). \tag{3.23}$$

Plots are presented in Fig. 3.4 for the SBJ, where it is seen that the lift-to-drag ratio has a maximum with respect to the lift coefficient and with respect to the velocity. Since $E = L/D$ and the lift is held constant, the velocity for maximum lift-to-drag ratio is identical with the velocity for minimum drag.

3.4   Aerodynamics: Prediction

This section presents an algorithm for calculating the aerodynamic characteristics of a subsonic jet airplane. The two quantities of interest are the angle of attack $\alpha = \alpha(C_L, M)$ and the drag polar $C_D = C_D(C_L, M, R_e)$.

3.5   Angle of Attack

The lift of an airplane is the lift of the wing-body combination plus the lift of the horizontal tail. The lift of a wing-body combination is a

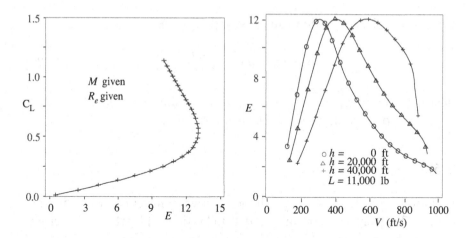

Figure 3.4: Lift-to-Drag Ratio (SBJ)

complicated affair in that the body produces some lift and interference effects between the wing and the body increase the lift of the body. It has been observed that the lift of a wing-body combination can be replaced by the lift of the entire wing (including that portion which passes through the fuselage). The lift of the horizontal tail is neglected with respect to that of the wing. Hence, the lift of the airplane is approximated by the lift of the entire wing. Geometrically, the wing is defined by its planform shape, its airfoil shapes along the span, and the shape of its chord surface. The only wings considered here are those with a straight-tapered planform shape, the same airfoil shape along the span, and a planar chord surface (no bend or twist). If a wing does not meet these conditions, it can be replaced by an average wing that does. For example, if the airfoil has a higher thickness ratio at the root than it does at the tip, an average thickness can be used. The aerodynamic characteristics of airfoils and wings have been taken from Refs. AD and Ho.

Over the range of lift coefficients where aircraft normally operate, the lift coefficient of the wing can be assumed to be linear in the angle of attack (see Fig. 3.2), that is,

$$C_L = C_{L_\alpha}(\alpha - \alpha_{0L}) \tag{3.24}$$

where α_{0L} is the *zero-lift angle of attack* and C_{L_α} is the *lift-curve slope* of the wing. This equation can be solved for α as

$$\alpha = \alpha_{0L}(M) + \frac{C_L}{C_{L_\alpha}(M)}. \tag{3.25}$$

Hence, to obtain α, it is necessary to determine α_{0L} and C_{L_α}. First, airfoils are discussed, then wings, then airplanes.

3.5.1 Airfoils

An airfoil is the cross-sectional shape of a two-dimensional wing (infinite span). The geometry of a cambered airfoil is defined by the geometries of the *basic symmetric airfoil* and the *camber line* shown in Fig. 3.5 where t is the maximum thickness and c is the chord. The basic symmetric airfoil

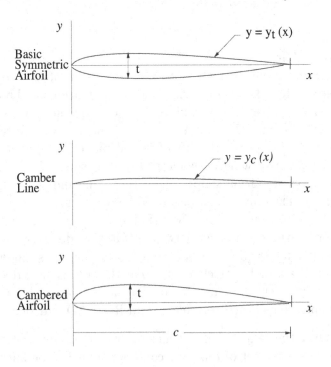

Figure 3.5: Airfoil Geometric Characteristics

shape $y = y_t(x)$ involves a number of parameters which are related to such geometric quantities as leading edge radius, trailing edge angle, thickness ratio (t/c), location of maximum thickness, etc. The camber line shape $y = y_c(x)$ involves parameters which are related to chord, leading edge slope, maximum displacement of the camber line from the chord and its location, etc. The basic symmetric airfoil and the camber line are combined to form the cambered airfoil.

An airfoil at an angle of attack α experiences a resultant aerodynamic force, and the point on the chord through which the line of action passes is called the *center of pressure* (Fig. 3.6). The resultant aerodynamic force is resolved into components parallel and perpendicular to the velocity vector called the drag and the lift.

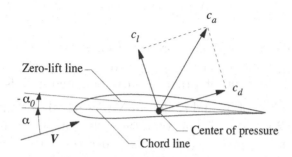

Figure 3.6: Center of Pressure

In general, the lift coefficient and the drag coefficient behave as

$$c_l = c_l(\alpha, M) \ , \quad c_d = c_d(\alpha, M, R_e) \tag{3.26}$$

where M is the Mach number and R_e is the Reynolds number. With lift coefficient as a variable instead of the angle of attack, these relations become

$$\alpha = \alpha(c_l, M) \ , \quad c_d = c_d(c_l, M, R_e). \tag{3.27}$$

By holding M and R_e constant and varying the lift coefficient, the above quantities vary as in Fig. 3.7 for a cambered airfoil. Here, α_0 is the zero-lift angle of attack, c_{l_α} is the lift-curve slope, and c_{l_i} is the ideal lift coefficient or the lift coefficient where the drag coefficient is a minimum. From this figure, it is seen that the angle of attack varies linearly with the lift coefficient and the drag coefficient varies quadratically with c_l over a wide range of c_l.

A systematic study of airfoil aerodynamics has been conducted by NACA. Aerodynamic data has been collected for various sets of thickness distributions and camber lines and are designated as NACA X-digit

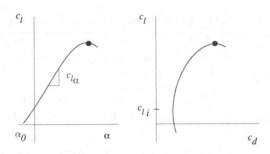

Figure 3.7: Airfoil Aerodynamic Characteristics - M, R_e given

series. The data are usually presented for $M = 0.2$, $R_e = 9 \times 10^6$, and means for making Mach number and Reynolds number corrections are available. Collections of airfoil data can be found in Refs. AD and Ho.

Values of α_0 and c_{l_α} for an NACA 64-109 airfoil are presented in Table 3.2 along with how they vary with the Mach number. While the numbers are presented in terms of degrees, all of the formulas use the numbers in radians. Each of the numbers in the designation 64-109 means something. The 6 denotes a 6-series airfoil which indicates a particular geometry for the thickness distribution and the camber line. The 4 means that the *peak suction* (minimum pressure) occurs at $x_{ps}/c = 0.4$; the 1 means that the *ideal lift coefficient* (lift coefficient for minimum drag coefficient) is $c_{l_i} = 0.1$; and the 09 means that the *thickness ratio* is $t/c = 0.09$.

Table 3.2: Data for the NACA 64-109 Airfoil

Parameter	$M = 0.2$ (M=0)	Variation with Mach number
α_0	-0.5 deg	Negligible
c_{l_α}	0.110 deg^{-1}	$c_{l_\alpha} = \frac{(c_{l_\alpha})_{M=0}}{\sqrt{1-M^2}}$

3.5.2 Wings and horizontal tails

For a straight-tapered wing or horizontal tail, the basic geometric properties of half the wing (Fig. 3.8), which is a trapezoid, are the *root chord* c_r, the *tip chord* c_t, the *semi-span* $b/2$, and the *sweep* of the quarter-chord line (m=.25). Note that $m = 0$ is the leading edge, and $m = 1$ is the trailing edge. Given values for these quantities, derived quantities

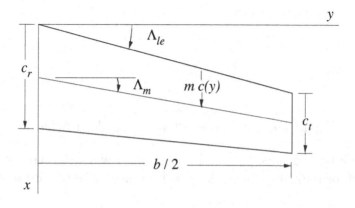

Figure 3.8: Geometry of a Straight, Tapered Wing

for the whole wing such as *planform area S, aspect ratio A, taper ratio* λ, sweep Λ_n of the n chord line $nc(y)$, and *mean aerodynamic chord* can be obtained from the following relations:

$$S = 2(\tfrac{b}{2})(\tfrac{c_r+c_t}{2})$$

$$A = \tfrac{b^2}{S}$$

$$\lambda = \tfrac{c_t}{c_r} \tag{3.28}$$

$$\tan\Lambda_n = \tan\Lambda_m - \tfrac{4}{A}(n-m)\tfrac{1-\lambda}{1+\lambda}$$

$$\bar{c} = \tfrac{2c_r}{3}\tfrac{1+\lambda+\lambda^2}{1+\lambda}.$$

The mean aerodynamic chord is the chord of the equivalent rectangular wing. It has the same lift and the same pitching moment about the y-axis as the original wing.

For such a wing, α is defined as the angle of attack of the root airfoil which here is the same as the angle of attack of the wing chord plane. Assuming that the lift coefficient is linear in the angle of attack (Fig. 3.9), two important parameters are the zero-lift angle of attack α_{0L}

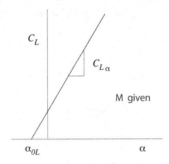

Figure 3.9: Lift Coefficient versus Angle of Attack

and the lift-curve slope C_{L_α}. For the assumed wing, the zero-lift angle of attack of the wing equals the zero-lift angle of attack of the airfoil, that is,

$$\alpha_{0L} = \alpha_0. \tag{3.29}$$

An accepted formula for the lift-curve slope of a swept wing is the following:

$$C_{L_\alpha} = \frac{\pi A}{1 + \sqrt{1 + (A/2\kappa)^2[1 + \tan^2 \Lambda_{hc} - M^2]}} \tag{3.30}$$

where Λ_{hc} is the half-chord sweep angle and where

$$\kappa = \frac{(c_{l_\alpha})_{M=0}}{(c_{l_\alpha})_{theory}}. \quad (c_{l_\alpha})_{theory} = 6.28 + 4.7(t/c) \tag{3.31}$$

is the ratio of the airfoil lift-curve slope to the theoretical value. From Table 3.2, it is seen that $\kappa = .94$ for the NACA 64-109. It is close to unity, as it is for most airfoils.

3.5.3   Airplanes

The angle of attack of an entire airplane is the angle between the x_b axis and the velocity vector (see Fig. 3.10). It is recalled that the x_b axis

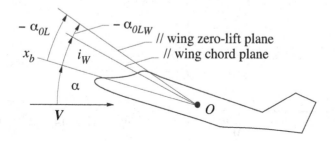

Figure 3.10: Airplane Angle of Attack

passes through the airplane center of gravity, and its orientation relative to the airplane depends on the type of airplane. For a passenger airplane, the x_b axis is usually parallel to the cabin floor. The wing is attached to the fuselage at an *incidence* i_W, the angle between the *wing chord plane* and the x_b axis. For the wing to produce zero lift, the airplane must be at the zero-lift angle of attack

$$-\alpha_{0L} = i_W - \alpha_{0LW} = i_W - \alpha_0 \tag{3.32}$$

which is independent of the Mach number. Since all of the lift of the airplane has been assumed to be produced by the wing, the airplane lift-curve slope is given by Eq. (3.30).

The quantities α_{0L} and $C_{L\alpha}$, which are needed to determine the angle of attack (3.25), have been computed for the SBJ (App. A) and are presented in Fig. 3.11.

3.6 Drag Coefficient

The subsonic flow field over an airplane consists of a thin viscous boundary layer close to the surface of the airplane, an inviscid flow region outside the boundary layer whose pressure distribution at the outside of the boundary layer is transmitted through the boundary layer to the surface

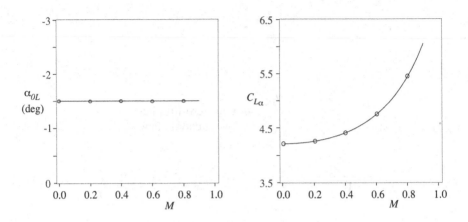

Figure 3.11: Zero-Lift Angle of Attack and Lift-Curve Slope (SBJ)

of the airplane, and if the speed is high enough embedded shock waves. Hence, to estimate the drag polar, the drag coefficient is divided into the *friction drag coefficient*, the *wave drag coefficient*, and the *induced drag coefficient*, that is,

$$C_D = C_{D_f}(M, R_e) + C_{D_w}(C_L, M) + C_{D_i}(C_L). \tag{3.33}$$

3.6.1 Friction drag coefficient

The friction drag coefficient is computed using the *equivalent parasite area method*. It is written as

$$C_{D_f} = \frac{f}{S} \tag{3.34}$$

where f is the total equivalent parasite area and S is the wing planform area. The former is the sum of the equivalent parasite area of each airplane component multiplied by 1.1 to account for miscellaneous contributions. Hence,

$$C_{D_f} = \frac{1.1}{S} \sum f_k \tag{3.35}$$

The equivalent parasite area is computed from the relation

$$f_k = C_{f_k} \, CF_k \, IF_k \, FF_k \, S_{wet_k} \tag{3.36}$$

where each term is defined below.

The average skin friction coefficient C_{f_k} is computed from the equation for the *skin friction coefficient* for turbulent flow over a flat plate, that is,

$$C_{f_k} = \frac{0.455}{(\log_{10} R_{e_k})^{2.58}} \tag{3.37}$$

where the freestream Reynolds number is given by

$$R_{e_k} = \frac{\rho V l_k}{\mu} \tag{3.38}$$

The reference length l_k of each component is defined in Table 3.3, where l denotes a body length and $\bar{c}$ denotes a mean aerodynamic chord.

The *compressibility factor* CF_k modifies the skin friction coefficient to account for Mach number effects and is given by

$$CF_k = (1.0 + 0.2M^2)^{-0.467} \tag{3.39}$$

The *interference factor* IF_k accounts for interference effects between two components, and representative values are tabulated in Table 3.3.

The *form factor* FF_k accounts for thickness effects. Formulas for its calculation are given by

$$
\begin{aligned}
FF_W &= 1.0 + 1.6(t/c)_W + 100(t/c)_W^4 \\
FF_H &= 1.0 + 1.6(t/c)_H + 100(t/c)_H^4 \\
FF_V &= 1.0 + 1.6(t/c)_V + 100(t/c)_V^4 \\
FF_B &= 1.0 + 60/(l/d)_B^3 + 0.0025(l/d)_B \\
FF_T &= 1.0 + 60/(l/d)_T^3 + 0.0025(l/d)_T \\
FF_N &= 1.0 + 0.35/(l/d)_N
\end{aligned}
\tag{3.40}
$$

where t/c is the airfoil *thickness ratio* (maximum thickness divided by chord) and l/d is the body *fineness ratio* (length divided by maximum diameter).

Table 3.3: Reference Lengths and Interference Factors

Airplane Component	k	Reference Length	Interference Factor
Body	B	l_B	1.20
Wing	W	$\bar{c}_W$	1.20
Horizontal Tail	H	$\bar{c}_H$	1.10
Vertical Tail	V	$\bar{c}_V$	1.10
Wing Nacelles	N	l_N	1.30
Body Nacelles	N	l_N	1.50
Tip Tanks	T	l_T	1.25

Finally, the *wetted area* S_{wet_k} is the external surface area of the airplane touched by the air and is computed using simple geometric shapes. Lifting surfaces can be assumed to be flat plates; bodies can be made up of cones and cylinders; and nacelles can be represented by open-ended cylinders.

The total equivalent parasite area is the sum of the component equivalent parasite areas plus ten percent of this total to account for miscellaneous contributions.

3.6.2   Wave drag coefficient

The rapid increase in the drag coefficient due to the formation of shock waves at high speeds is called wave drag. Shock waves cause the boundary layer to separate, thus increasing the drag. Wave drag begins at the *Mach number for drag divergence* which is shown in Fig. 3.12. Because a subsonic airplane will not fly above M_D because the drag is too high, an approximate formula is used to represent the wave drag coefficient and is given by

$$C_{D_w} = 29.2(M - M_D)^3 \ , \quad M \geq M_D \tag{3.41}$$

where the drag divergence Mach number M_D is computed from the wing and airfoil geometry as follows:

$$M_D = g_1 - g_2 C_L \tag{3.42}$$

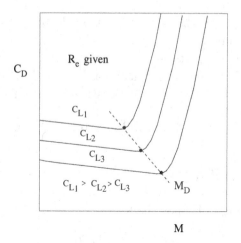

Figure 3.12: Mach Number for Drag Divergence

The quantities g_1 and g_2 are defined in terms of wing and airfoil properties as

$$
\begin{aligned}
g_1 &= [1 + 0.189(4\Lambda_{ps} - 3\Lambda_{mt})][1 - 1.4(t/c)_W \\
&\quad -0.06(1 - x_{ps}/c)] - 0.0368 \\
g_2 &= 0.33(0.65 - x_{ps}/c)[1 + 0.189(4\Lambda_{ps} - 3\Lambda_{mt})]
\end{aligned}
\tag{3.43}
$$

where x_{mt}/c is the chordwise location of the airfoil maximum thickness, Λ_{mt} is the sweep of the maximum thickness line, x_{ps}/c is the airfoil peak suction location, and Λ_{ps} is the sweep of the peak suction line. The wave drag coefficient is included only if the free-stream Mach number is greater than the drag divergence Mach number. If it is less than the drag divergence Mach number, then the wave drag coefficient is equal to zero.

3.6.3 Induced drag coefficient

Induced drag is caused by the rotational flow about the tip vortices. It is called vortex drag, drag due to lift, or induced drag. The corresponding drag coefficient is given by

$$
C_{D_i} = \frac{C_L^2}{\pi A_W e (1 + 0.5 d_T/b_W)}.
\tag{3.44}
$$

The induced drag of the horizontal tail is neglected. The wing tip tank diameter d_T accounts for the tendency of tip tanks to reduce the induced drag because of an end plate effect. *Oswald's efficiency factor e* accounts for the difference between an elliptical planform and the straight-tapered planform. It is estimated from the statistical equation

$$e = (1 - 0.045 A_W^{0.68})(1 - 0.227 \Lambda_{qcw}^{1.615}). \tag{3.45}$$

3.6.4   Drag polar

In general, the drag polar has the form

$$C_D = C_D(C_L, M, R_e). \tag{3.46}$$

The drag polar of the SBJ is shown in Fig. 3.13 where $R_e' \triangleq \rho V/\mu$ is the Reynolds number per ft. Values of C_D are shown for $R_e' = 1.0 \times 10^6$ and several values of the Mach number. It is seen that for $M \leq .8$ the Mach number has little effect on the drag polar. In Fig. 3.14 the effect of R_e' on C_D is shown for $M \leq 0.8$. For $C_L = 0.3$, the values of C_D are .0253, .0265, and .0287. The change in C_D relative to the middle value is 5% in one direction and 8% in the other. Hence, the change in the Reynolds number has only a small effect on the value of C_D. If the Reynolds number which is selected for the computation is around the lowest value expected to be experienced in flight, the drag coefficient will be on the conservative side (slightly higher). The value $R_e' = 1.0 \times 10^6$ is used for the subsequent calculations. Hence, if the effects of Reynolds number changes are neglected, the functional relation for the drag coefficient becomes

$$C_D = C_D(C_L, M). \tag{3.47}$$

It should be mentioned that drag polars can be created for take-off, climb, cruise, and landing by changing the value of the Reynolds number which is used to compute the polar.

3.7   Parabolic Drag Polar

As stated previously, the angle of attack can be assumed to be linear in the lift coefficient over a wide range of values of C_L, but not near the maximum. This relationship is given by Eq. (3.25). Over the same range

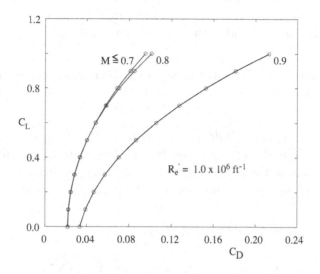

Figure 3.13: Drag Polar for the SBJ, R_e given

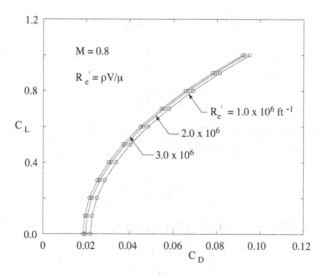

Figure 3.14: Drag Polar for the SBJ, M given

of C_L, the drag coefficient can be assumed to be a parabolic function of the lift coefficient. The lift coefficient range for which these approximations are valid includes the values of C_L normally experienced in flight.

One form of the *parabolic drag polar* is given by (see Prob. 3.6 for another)

$$C_D = C_{D_0}(M) + K(M)C_L^2 \tag{3.48}$$

where C_{D_0} is the *zero-lift drag coefficient*, KC_L^2 is the *induced drag coefficient*, and K is the *induced drag factor*. To obtain the parabolic drag polar for a given Mach number, the actual polar is fit by the parabola (3.48) using least squares and forcing the parabola to pass through the point where $C_L = 0$.

Thus, the equations for C_{D_0} and K are given by

$$C_{D_0} = (C_D)_{C_L=0}, \ \ K = \sum_{k=1}^{n}(C_{D_k} - C_{D_0})C_{L_k}^2 / \sum_{k=1}^{n} C_{L_k}^4 \tag{3.49}$$

where n is the number of points being fit. On the other hand, if $M < M_D$,

$$C_{D_0} = C_{D_f}$$
$$K = \frac{1}{\pi A_W e (1 + 0.5 d_T/b_W)}. \tag{3.50}$$

For the parabolic drag polar, the lift-to-drag ratio is given by

$$E = \frac{C_L}{C_{D_0} + KC_L^2}. \tag{3.51}$$

For a given Mach number (C_{D_0} and K constant), E has a maximum when the lift coefficient has the value

$$C_L^* = \sqrt{\frac{C_{D_0}}{K}}. \tag{3.52}$$

The maximum lift-to-drag ratio then given by

$$E^* = \frac{1}{2\sqrt{C_{D_0}K}}. \tag{3.53}$$

The quantities C_{D_0}, K, C_L^* and E^* have been computed for the SBJ (App. A) with a $R_e' = 1.0 \times 10^6$ ft^{-1}. The curves are presented in

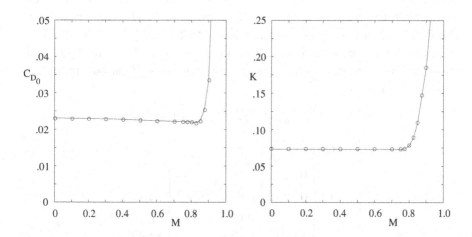

Figure 3.15: C_{D_0} and K for the SBJ

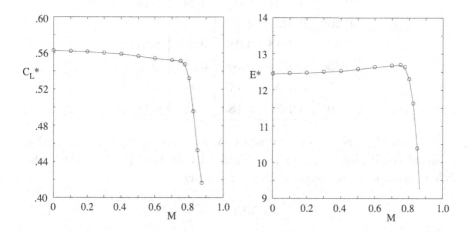

Figure 3.16: C_L^* and E^* for the SBJ

Figs. 3.15 and 3.16, and the values are given in Table 3.4. It is observed from these figures that C_{D_0}, K, C_L^*, and E^* are nearly constant for Mach numbers in the range $0 < M < .8$. This range of Mach numbers defines the flow regime known as subsonic flow. In the neighborhood of M = .8, shock waves begin to form in the flow field around the airplane, and the aerodynamic parameters undergo large changes. This flow regime is called transonic flow.

Table 3.4: Parabolic Drag Polar Data (SBJ)

M	C_{D_0}	K	C_{L^*}	E^*
0.0	0.0231	0.073	0.563	12.18
0.1	0.0231	0.073	0.562	12.19
0.2	0.0230	0.073	0.561	12.21
0.3	0.0229	0.073	0.560	12.23
0.4	0.0228	0.073	0.559	12.26
0.5	0.0226	0.073	0.556	12.31
0.6	0.0224	0.073	0.554	12.37
0.7	0.0223	0.073	0.552	12.41
0.750	0.0222	0.073	0.551	12.43
0.775	0.0221	0.074	0.547	12.36
0.800	0.0221	0.078	0.532	12.04
0.825	0.0218	0.089	0.495	11.36
0.850	0.0224	0.109	0.452	10.12
0.875	0.0255	0.147	0.416	8.173
0.9	0.0336	0.185	0.426	6.345

If C_{D_0} and K are assumed constant (parabolic drag polar with constant coefficients), Eq. (3.48) combined with Eqs. (3.14) leads to the following dimensional expression for the drag:

$$D = \frac{1}{2}C_{D_0}\rho SV^2 + \frac{2KL^2}{\rho SV^2}. \tag{3.54}$$

Note that $D = D(h, V, L)$. If $L = W$ and if the altitude and the weight are given, the drag has a minimum with respect to the velocity when

$$V = \sqrt{\frac{2W}{\rho S}}\sqrt{\frac{K}{C_{D_0}}} = \sqrt{\frac{2W}{\rho S C_L^*}} \triangleq V^*. \tag{3.55}$$

At this velocity, the drag has the minimum value

$$D^* = 2\sqrt{C_{D_0}K}\, W = \frac{W}{E^*}. \tag{3.56}$$

For a given altitude and lift, minimizing the drag is equivalent to maximizing the lift-to-drag ratio. This explains why the velocity for minimum drag equals the velocity for maximum lift-to-drag ratio.

3.8 Propulsion: Thrust and SFC

In this section, the propulsion characteristics appearing in the equations of motion, thrust and specific fuel consumption (SFC), are discussed functionally for two subsonic airbreathing jet engines, a turbojet and a turbofan. Next, approximate expressions are presented for the thrust and specific fuel consumption. The point of view taken here is that engine data is available from the engine manufacturer so that no formulas are presented for estimating these quantities.

3.8.1 Functional relations

One manner of presenting engine data is in terms of *corrected thrust* and *corrected specific fuel consumption*, that is,

$$
\begin{aligned}
T_c &= \tfrac{T}{\delta} \\
C_c &= \tfrac{C}{\sqrt{\theta}}
\end{aligned}
\tag{3.57}
$$

where the dimension of thrust is lb and that of specific fuel consumption is l/hr. The pressure ratio δ and temperature ratio θ are defined as

$$
\begin{aligned}
\delta &= \tfrac{\bar{p}}{p_s} \\
\theta &= \tfrac{\bar{\tau}}{\tau_s}
\end{aligned}
\tag{3.58}
$$

where the sea level static pressure p_s is 2116.2 lb/ft^2 and the sea level static temperature τ_s is 518.69 °R. The total pressure $\bar{p}$ and total temperature $\bar{\tau}$ for isentropic flow of air (ratio of specific heats = 1.4) can be expressed as

$$
\begin{aligned}
\bar{p} &= p(1 + 0.2M^2)^{3.5} \\
\bar{\tau} &= \tau(1 + 0.2M^2)
\end{aligned}
\tag{3.59}
$$

Engine manufacturer's data (see Fig. 3.17 for a turbojet or Fig. 3.18 for a turbofan) shows that the corrected thrust and specific fuel consumption satisfy functional relations of the form

$$T_c = T_c(M, \eta), \quad C_c = C_c(M, \eta). \tag{3.60}$$

The *corrected engine speed* η is related to the power setting $P = N/N_{max}$, where N is the engine rpm, as follows:

$$\eta = P\eta_{max}. \tag{3.61}$$

For a turbojet, the value of η_{max} is given by

$$\eta_{max} = \min(1.05, 1.0/\sqrt{\theta}) \tag{3.62}$$

For the turbofan, the value of η_{max} is

$$\eta_{max} = \begin{cases} (1,958 + 47h/5,000)/\bar{\tau}, & h \le 5,000 \text{ ft} \\ 2,005/\bar{\tau}, & h > 5,000 \text{ ft} \end{cases} \tag{3.63}$$

With regard to the power setting, the following are accepted definitions:

$$\begin{aligned} P &= 1.00, \quad \text{take--off thrust} \\ P &= 0.98, \quad \text{maximum continuous thrust} \end{aligned} \tag{3.64}$$

By combining Eqs. (3.57) through (3.63), it is seen that the thrust and specific fuel consumption satisfy functional relations of the form

$$T = T(h, V, P), \quad C = C(h, V, P) \tag{3.65}$$

These are the functional relations used earlier to discuss the equations of motion.

Fig. 3.17 shows the corrected thrust and the corrected specific fuel consumption for a turbojet. The numbers for these figures are given in Table 3.6. Note that the maximum sea level static thrust (h=0, M=0, $\eta_{max} = 1$) is around 3,000 lb. Fig. 3.18 gives the corrected thrust and the corrected specific fuel consumption for a turbofan. The numbers for these figures are given in Table 3.6. The maximum sea level static thrust (h=0, M=0. $\eta_{max}=3.8$) is around 4,000.

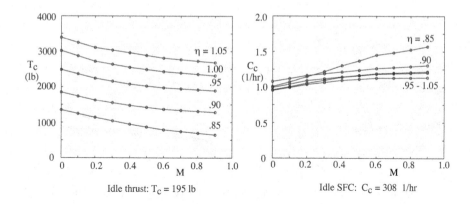

Idle thrust: $T_c = 195$ lb   Idle SFC: $C_c = 308$ 1/hr

Figure 3.17: Corrected Thrust and SFC: GE CJ610-6 Turbojet

Table 3.5: Turbojet Engine Data

	$\eta = .85$		$\eta = .90$		$\eta = .95$		$\eta = 1.00$		$\eta = 1.05$	
M	T_c	C_c	T_c	C_c	T_c	C_c	T_c	C_c	T_c	C_c
0.0	1,353	1.009	1853	.9601	2,488	0.954	3022	.9954	3,390	1.080
0.1	1,245	1.071	1740	1.004	2,365	0.993	2871	1.041	3,248	1.119
0.2	1,138	1.146	1626	1.054	2,242	1.037	2720	1.092	3,105	1.163
0.3	1,042	1.216	1551	1.094	2.160	1.065	2635	1.117	3.033	1.187
0.4	945	1.302	1475	1.139	2,078	1.095	2550	1.143	2,960	1.213
0.5	867	1.369	1420	1.160	2,026	1.111	2490	1.166	2,885	1.237
0.6	789	1.449	1365	1.184	1,974	1.127	2430	1.189	2,810	1.262
0.7	742	1.484	1337	1.190	1,947	1.129	2392	1.199	2,768	1.276
0.8	695	1.524	1309	1.196	1,921	1.130	2355	1.208	2,725	1.290
0.9	647	1.569	1281	1.203	1,894	1.132	2318	1.218	2,683	1.305

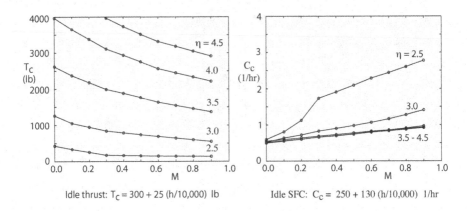

Figure 3.18: Corrected Thrust and SFC: Garrett TFE 731-2 Turbofan

Table 3.6: Turbofan Engine Data

M	$\eta = 2.5$		$\eta = 3.0$		$\eta = 3.5$		$\eta = 4.0$		$\eta = 4.5$	
	T_c	C_c	T_c	C_c	T_c	C_c	T_c	C_c	T_c	C_c
0.0	427	0.598	1,260	0.520	2,618	0.501	3,962	0.516	4,929	0.553
0.1	333	0.808	1,050	0.637	2,373	0.553	3,651	0.560	4,583	0.595
0.2	253	1.123	945	0.721	2,182	0.602	3,379	0.605	4,274	0.638
0.3	172	1.735	840	0.827	1,992	0.660	3,018	0.657	3,966	0.688
0.4	164	1.909	786	0.900	1,884	0.699	2,934	0.696	3,731	0.731
0.5	156	2.095	736	0.980	1,765	0.747	2,753	0.742	3,514	0.776
0.6	149	2.291	690	1.065	1,635	0.807	2,566	0.797	3,315	0.823
0.7	145	2.445	643	1.163	1,546	0.855	2,449	0.834	3,182	0.857
0.8	142	2.607	596	1.277	1,457	0.909	2,333	0.876	3,047	0.895
0.9	138	2.778	550	1.409	1,367	0.969	2,218	0.922	2,914	0.936

3.8.2   Approximate formulas

Approximate formulas have been proposed for thrust and specific fuel consumption, that is,

$$T = T_t(V, P)(\rho/\rho_t)^a, \quad C = C_t(V, P)(\rho/\rho_t)^b \qquad (3.66)$$

where the subscript t denotes the tropopause. The main features of these formulas are that they are exact in the constant temperature part of the stratosphere ($a = 1$, $b = 0$) and that they are extended down into the troposphere by putting an arbitrary exponent on the density ratio. These formulas are more valid near the tropopause than they are near sea level.

The fitting of Eqs. (3.66) to engine data in the troposphere has been carried out for the GE turbojet. The quantities $T_t(V, P)$ and $C_t(V, P)$ are obtained from the data of Fig. 3.17 at the tropopause and are presented in Fig. 3.19. Next, Eqs. (3.66) are solved for a and b and plotted in Fig. 3.20. The values of a and b are taken from these figures to be $a = 1.2$ and $b = 0.1$.

In the interest of developing analytical performance results, it is observed from Fig. 3.19 that T_t can be assumed to be independent of the Mach number and that C_t can be assumed to be independent of both the Mach number and the power setting. Hence, the T_t and C_t in the above approximations can be rewritten as

$$T_t = T_t(P), \quad C_t = \text{Const.} \qquad (3.67)$$

Values of $T_t(P)$ and C_t are given in Table 3.7.

Table 3.7: Tropopause Thrust and SFC - Turbojet

P	$T_t(P)$	C_t
	lb	hr^{-1}
.83	285	1.18
.88	450	1.18
.93	600	1.18
.98	710	1.18

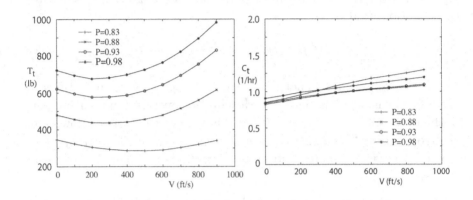

Figure 3.19: Tropopause Thrust and SFC: GE CJ610-6 Turbojet

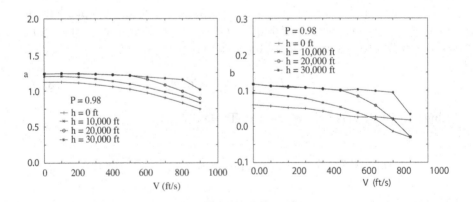

Figure 3.20: Thrust and SFC Exponents: GE CJ610-6 Turbojet

The fitting of Eqs. (3.66) to engine data in the troposphere has also been carried out for the Garrett turbofan. The quantities $T_t(V, P)$ and $C_t(V, P)$ are obtained from the data of Fig. 3.18 at the tropopause and are presented in Fig. 3.21. Next, Eqs. (3.66) are solved for a and b and plotted in Fig. 3.22. The values of a and b are taken from these figures to be $a = 1.0$ and $b = 0.0$. Note that they are the same as the statosphere values.

In the interest of developing analytical performance results, it is observed from Figs. 3.21 that T_t can be assumed to be independent of the Mach number and that C_t can be assumed to be independent of both the Mach number and the power setting. Hence, the T_t and C_t in the above approximations can be rewritten as Eqs. (3.67). Values of $T_t(P)$ and C_t are given in Table 3.8.

Table 3.8: Tropopause Thrust and SFC - Turbofan

P	$T_t(P)$	C_t
	lb	hr^{-1}
.83	973	.725
.88	1067	.725
.93	1095	.725
.98	1112	.725

3.9 Ideal Subsonic Airplane

The *Ideal Subsonic Airplane* (ISA) is defined as an airplane that has a parabolic drag polar with constant coefficients, a thrust independent of the velocity, and a specific fuel consumption independent of the velocity and the power setting. Hence, the drag is given by

$$D = \frac{1}{2}C_{D_0}\rho SV^2 + \frac{2KL^2}{\rho SV^2} \tag{3.68}$$

and the thrust and specific fuel consumption satisfy the relations

$$T = T_t(P)(\rho/\rho_t)^a, \quad C = C_t(\rho/\rho_t)^b. \tag{3.69}$$

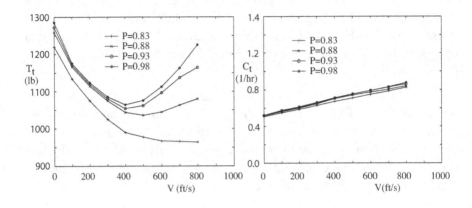

Figure 3.21: Tropopause Thrust and SFC: Garrett TFE 731-2 Turbofan

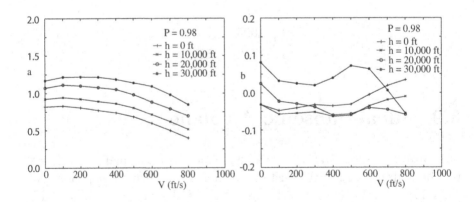

Figure 3.22: Thrust and SFC Exponents: Garrett TFE 731-2 Turbofan

Note that D, T, and C satisfy the functional relations

$$D = D(h, V, L) , \quad T = T(h, P) , \quad C = C(h). \qquad (3.70)$$

The aerodynamics of the *Ideal SBJ* (ISBJ, App. A) are given by

$$C_{D_0} = 0.023, \quad K = 0.073. \qquad (3.71)$$

The propulsion characteristics of the GE turbojet are given by

$$\text{Troposphere}: a = 1.2, \ b = 0.1 \qquad (3.72)$$

$$\text{Stratosphere}: a = 1, \ b = 0, \qquad (3.73)$$

with $T_t(P)$ and C_t given in Table 3.7.

The propulsion characteristics of the Garrett turbofan are given by

$$\text{Troposphere}: a = 1, \ b = 0 \qquad (3.74)$$

$$\text{Stratosphere}: a = 1, \ b = 0, \qquad (3.75)$$

with $T_t(P)$ and C_t given in Table 3.8. It is not possible to just replace the GE turbojet in the ISBJ by the Garrett turbofan. Each turbofan engine weighs about 300 lb more than the turbojet.

Problems

The answers to the problems involving the computation of the SBJ geometry or aerodynamics are given in App. A to help keep you on track. Once you have completed an assignment, you should use the numbers given in App. A instead of those you have calculated.

3.1 Perform the tasks listed below for the wing of the SBJ (App. A). The wing planform extends from the fuselage centerline to the outside of the tip tanks.

Starting from the measured values for the root chord, the tip chord, the span, and the sweep angle of the quarter chord line (App. A) calculate the planform area, the aspect ratio, the taper ratio, and the sweep of the leading edge. Also, calculate the length of the mean aerodynamic chord.

3.2 To demonstrate that Eq. (3.30) reduces to known incompressible thin-wing results, show that, for $M = 0$, $\Lambda_{hc} = 0$, and $\kappa = 1$,

$$C_{L_\alpha} = \frac{\pi A}{1 + \sqrt{1 + (A/2)^2}}.$$

Then, for large aspect ratio wings $[(A/2)^2 >> 1]$, show that

$$C_{L_\alpha} = \frac{2\pi A}{2 + A}.$$

Finally, for a two-dimensional wing (airfoil) for which $A = \infty$, show that

$$C_{L_\alpha} = 2\pi$$

which is the theoretical thin-airfoil result.

3.3 Prove that the surface area of a right circular cone (excluding the base) is given by

$$A = \pi R \sqrt{R^2 + h^2}$$

where R denotes the base radius and h denotes the height.

3.4 Calculate the wetted area of each component of the SBJ (App. A).

3.5 Assume that the SBJ (App. A) is operating in level flight ($L = W$) at $h = 30{,}000$ ft, $M = 0.7$, and $W = 11{,}000$ lb. The lift coefficient is given by $C_L = 2W/\rho S V^2$.

   a. Compute the Mach number for drag divergence.

   b. Calculate C_{D_0} and K for this flight condition. In doing this calculation, remember that there are two nacelles and two tip tanks.

3.6 Another form of the parabolic drag polar is given by

$$C_D = C_{D_m}(M) + K_m(M)[C_L - C_{L_m}(M)]^2$$

where C_{D_m}, C_{L_m} define the minimum drag point. If the Mach number is given, find the lift coefficient for maximum lift-to-drag ratio and the maximum lift-to-drag ratio for this polar.

Chapter 4

Cruise and Climb of an Arbitrary Airplane

In the mission segments known as cruise and climb, the accelerations of airplanes such as jet transports and business jets are relatively small. Hence, these performance problems can be studied by neglecting the tangential acceleration $\dot{V}$ and the normal acceleration $V\dot{\gamma}$ in the equations of motion. Broadly, this chapter is concerned with methods for obtaining the distance and time in cruise and the distance, time, and fuel in climb for an arbitrary airplane. What is meant by an arbitrary airplane is that cruise and climb performance is discussed in terms of the functional relations $D(h, V, L), T(h, V, P)$, and $C(h, V, P)$. These functional relations can represent a subsonic airplane powered by turbojet engines, a supersonic airplane powered by turbofan engines, and so on. In order to compute the cruise and climb performance of a particular airplane, the functional relations are replaced by computer functions which give the aerodynamic and propulsion characteristics of the airplane.

In Chap. 5, the theory of this chapter is applied to an Ideal SBJ. There, approximate analytical formulas are used for the aerodynamics and propulsion. Analytical solutions are obtained for the distance and time in cruise and the distance, time, and fuel in climb.

The airplane is a controllable dynamical system which means that the equations of motion contain more variables than equations, that is, one or more mathematical degrees of freedom (MDOF). In standard performance problems, the MDOF are reduced to one, and it is associated with the velocity profile flown by the airplane during a particular

mission leg. Then, the performance of an airplane can be computed for
a particular velocity profile, or trajectory optimization can be used to
find the optimal velocity profile. This is done by solving for the distance,
time, and fuel in terms of the unknown velocity profile. After selecting
distance, time or fuel as the performance index, optimization theory is
used to find the corresponding optimal velocity profile. An example is
to find the velocity profile which maximizes the distance in cruise from
one weight to another, that is, for a given amount of fuel. Besides being important all by themselves, optimal trajectories provide yardsticks
by which arbitrary trajectories (for example, constant velocity) can be
evaluated.

       This chapter begins with a discussion of how flight speeds are
represented and what some limitations on flight speed are.

4.1   Special Flight Speeds

In presenting the performance characteristics of an aiplane, several quantities can be used to represent the velocity. The velocity of the airplane
relative to the atmosphere, V, is called the *true airspeed*. If σ denotes
the ratio of the atmospheric density at the altitude the airplane is operating, ρ, to that at sea level, ρ_s, the *equivalent airspeed* is defined as
$V_e = \sqrt{\sigma}V$. Note that the equivalent airspeed is proportional to the
square root of the dynamic pressure $\bar{q} = \rho V^2/2$. This airspeed is important for low-speed flight because it can be measured mechanically.
The air data system measures the dynamic pressure and displays it to
the pilot as *indicated airspeed*. The indicated airspeed is the equivalent
airspeed corrupted by measurement and instrument errors. To display
the true airspeed to the pilot requires the measurement of static pressure, dynamic pressure, and static temperature. Then, the true airspeed
must be calculated. For high-speed flight, it has become conventional to
use *Mach number* as an indication of flight speed. High-speed airplanes
have a combined airspeed indicator and Mach meter.

       In this chapter, the performance of a subsonic business jet (the
SBJ of App. A) is computed to illustrate the procedures and results.
Because this airplane operates in both low-speed and high-speed regimes,
the true airspeed is used to present performance. Note that true airspeed
can be roughly converted to Mach number by using $a = 1,000$ ft/s. Other
conversion formulas are useful. If a statute mile has 5,280 ft, 1 mi/hr =

1.4667 ft/s. If a nautical mile has 6,076 ft, 1 kt = 1.6878 ft/s. A kt is one nautical mile per hour.

4.2 Flight Limitations

The lowest speed at which an airplane can maintain steady level flight (constant altitude) is called the stall speed. In steady level flight, the equation of motion normal to the flight path is given by

$$L = W \qquad (4.1)$$

where the component of the thrust has been neglected. Because of the definition of lift coefficient, this equation can be written as

$$\frac{1}{2}C_L\rho S_W V^2 = W \qquad (4.2)$$

and says that for a given weight the product $C_L V^2$ is constant. Hence, the lower the speed of the airplane is the higher the lift coefficient must be. Since there is a maximum lift coefficient, there is a minimum speed at which an airplane can be flown in steady level flight. This speed is called the *stall speed* and is given by

$$V_{stall} = \sqrt{\frac{2W}{\rho S_W C_{L_{max}}}}. \qquad (4.3)$$

Some airplanes are speed limited by their structure or control capability. This limit can take the form of a maximum dynamic pressure, so that

$$V_{\bar{q}max} = \sqrt{\frac{2\bar{q}_{max}}{\rho}} \qquad (4.4)$$

by definition of dynamic pressure. This limit can affect the best climb speed of the airplane as well as the maximum speed.

At high speeds, some airplanes may have a maximum operating Mach number. Hence, by definition of Mach number, the limiting speed is

$$V_{M_{max}} = M_{max}a \qquad (4.5)$$

While this speed decreases with altitude in the troposphere, it is constant in the constant temperature part of the stratosphere.

Note that the stall speed and the maximum dynamic pressure speed are actually constant equivalent airspeeds.

4.3   Trajectory Optimization

An important part of airplane performance is the optimization of trajectories. These problems can be solved by a branch of mathematics called Calculus of Variations, which in recent times is also called Optimal Control Theory. A simple problem which occurs frequently in airplane trajectory optimization is to find the curve $y(x)$ which maximizes the performance index

$$J = \int_{x_0}^{x_f} f(x, y) \, dx \qquad (4.6)$$

where x_0 and x_f are the initial and final values of the variable of integration. It is known that the curve $y(x)$ which maximizes the integral (4.6) satisfies the conditions

$$\left. \frac{\partial f}{\partial y} \right|_{x=Const} = 0, \qquad \left. \frac{\partial^2 f}{\partial y^2} \right|_{x=Const} < 0. \qquad (4.7)$$

The first condition gives the curve $y(x)$ which optimizes the integral, and the second condition identifies it as a maximum. For a minimum, < 0 is replaced by > 0. It is possible to verify the nature (maximum or minimum) of the optimal curve by looking at a plot of the function $f(x, y)$ versus y for several values of x.

Because of the simple form of Eq. (4.6), it is easy to verify Eq. (4.7). The integral is the area under the integrand. To maximize the integral, it is necessary to maximize the area. At each value of the variable of integration, the area is maximized by maximizing F with respect to y. The conditions for doing this are Eqs. (4.7).

4.4   Calculations

In this chapter, example calculations are made for the cruise and climb performance of the Subsonic Business Jet (SBJ) in App. A. The airplane is assumed to be operating in a constant gravity standard atmosphere, to have a parabolic drag polar $C_D = C_{D_0}(M) + K(M)C_L^2$, and to be powered by two GE CJ610-6 turbojets. In the calculation of atmospheric properties, drag, thrust, and specific fuel consumption, it is assumed that the altitude h, the velocity V, the weight W, and the power setting P are known. The calculations have been done on MATLAB, and three functions are needed.

The first function contains the standard atmosphere equations of Sec. 3.1. Given h, the function returns the atmospheric properties, particularly density $\rho(h)$ and speed of sound $a(h)$.

The second function computes the drag. With $M = V/a(h)$, Table 3.4 is interpolated for $C_{D_0}(M)$ and $K(M)$. Then, the lift coefficient is obtained from $C_L = 2W/\rho(h)SV^2$, since $L = W$, and the drag coefficient is computed from the parabolic drag polar. The drag is obtained from $D = (1/2)C_D\rho(h)SV^2$.

The third function calculates the thrust and specific fuel consumption of the turbojet. First, the corrected engine speed η is determined from Eq. (3.61). Then, the corrected thrust and the corrected specific fuel consumption are obtained from Table 3.5 by a two-dimensional interpolation in terms of M and η. Finally, T and C are computed using the formulas at the beginning of Sec. 3.8.1.

In conclusion, given $h, V, W,$ and P, it is possible to calculate $\rho(h)$, $a(h)$, $D(h, V, W)$, $T(h, V, P)$, and $C(h, V, P)$

4.5 Flight Envelope

By definition, the flight envelope is the region of the velocity-altitude plane where the airplane can maintain steady level flight. The dynamic equations of motion for steady level flight ($\dot{V} = \dot{\gamma} = \gamma = 0$) are obtained from Eq. (2.29) as

$$
\begin{aligned}
T(h, V, P) - D(h, V, L) &= 0 \\
L - W &= 0
\end{aligned}
\tag{4.8}
$$

Then, since $L = W$,

$$
T(h, V, P) - D(h, V, W) = 0.
\tag{4.9}
$$

If the altitude, weight, and power setting are given, this equation can be solved for the velocity.

In Fig. 4.1, the thrust and the drag are sketched versus the velocity. For a subsonic airplane, the drag has a single minimum whose value does not change with the altitude (see for example Fig. 3.3). Also, for a subsonic jet engine, the thrust does not vary greatly with the velocity, but it does change with the altitude, decreasing as the altitude increases. Fig. 4.1 shows that for some altitude, there exist two solutions of Eq. (4.9), a low-speed solution and a high-speed solution. As the altitude decreases (thrust increases), there is some altitude where the low-speed solution is the stall speed and at lower altitudes ceases to exist. As the altitude increases (thrust decreases), there is an altitude where there is only one solution, and above that altitude there are no solutions. The region of the velocity-altitude plane that contains all of the level flight solutions combined with whatever speed restrictions are imposed on the airplane, is called the *flight envelope*.

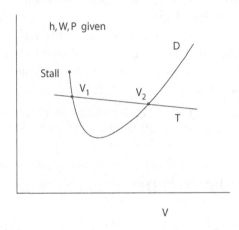

Figure 4.1: Thrust and Drag versus Velocity

The flight envelope has been computed for the SBJ weighing 11,000 lb and operating at maximum continuous thrust ($P=0.98$). It is shown in Fig. 4.2. The level flight solutions for this weight and power setting are indicated by $T - D = 0$. The stall speed ($C_{L_{max}}= 1.24$), the maximum dynamic pressure ($\bar{q}_{max}= 300$ lb/ft^2), and the maximum Mach number ($M_{max}=0.81$) limits are also plotted versus the altitude. The region enclosed by these curves is the flight envelope of the SBJ. The highest altitude at which the airplane can be flown in steady level flight is called the *ceiling* and is around 50,000 ft. Note that the highest speed at which the airplane can be flown is limited by the maximum dynamic pressure or the maximum Mach number.

Next to be discussed are the distance and time during cruise. These quantities have been called the *range* and *endurance*. However, because there is considerable distance and time associated with the climb, the range, for example, could be defined as the sum of the distance in climb and the distance in cruise.

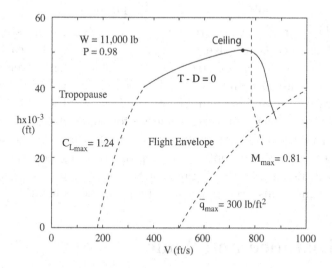

Figure 4.2: Flight Envelope of the SBJ

4.6   Quasi-steady Cruise

Of major importance in the mission profile is the cruise segment because airplanes are designed to carry a given payload a given distance. For a *constant altitude cruise*, the velocity vector is parallel to the ground, so that the equations of motion for quasi-steady level flight are given by Eqs. (2.29) with $\gamma = 0$, that is,

$$\dot{x} = V \tag{4.10}$$

$$0 = T(h, V, P) - D(h, V, L) \tag{4.11}$$

$$0 = L - W \tag{4.12}$$

$$\dot{W} = -C(h, V, P)T(h, V, P). \tag{4.13}$$

During a cruise, the altitude is constant and is not counted as a variable. Hence, these equations have two states, $x(t)$ and $W(t)$, three controls,

$V(t)$, $P(t)$, and $L(t)$, for a total of five variables. Since there are four equations, this system of equations has one mathematical degree of freedom, which is associated with the velocity profile $V(t)$.

The general procedure followed in studying quasi-steady airplane performance is to solve the equations of motion for each of the variables in terms of the unknown velocity profile. Then, given a velocity profile, the distance and the time for a given fuel can be determined. Since there are an infinite number of velocity profiles, it is desirable to find the one which optimizes some performance index. For cruise, there are two possible performance indices: distance (range) or time (endurance). Hence, the optimization problem is to find the velocity profile which maximizes the distance or the velocity profile which maximizes the time. This process is called trajectory optimization.

4.7   Distance and Time

To compute the distance and the time for a given amount of fuel (W_0 given, W_f given), Eqs. (4.11) and (4.12) require that $L = W$ and that

$$T(h, V, P) - D(h, V, W) = 0 \qquad (4.14)$$

This equation can be solved for the power setting as

$$P = P(h, V, W) . \qquad (4.15)$$

Next, the weight is made the variable of integration, and Eqs. (4.10) and (4.13) are rewritten as

$$
\begin{aligned}
-\frac{dx}{dW} &= \frac{V}{C(h,V,P)T(h,V,P)} \\
-\frac{dt}{dW} &= \frac{1}{C(h,V,P)T(h,V,P)}
\end{aligned}
\qquad (4.16)
$$

where all variables are now considered as functions of W. Then, Eq. (4.15) is used to eliminate the power setting so that

$$
\begin{aligned}
-\frac{dx}{dW} &= \frac{V}{C(h,V,P(h,V,W))T(h,V,P(h,V,W))} \triangleq F(W, V, h) \\
-\frac{dt}{dW} &= \frac{1}{C(h,V,P(h,V,W))T(h,V,P(h,V,W))} \triangleq G(W, V, h)
\end{aligned}
\qquad (4.17)
$$

where F and G are called the *distance factor* and the *time factor*. Since the altitude is constant and $V = V(W)$, the integration can be performed in principle to obtain $x(W)$ and $t(W)$ as follows:

$$x - x_0 \ = \ \int_W^{W_0} F(W, V, h) dW \tag{4.18}$$

$$t - t_0 \ = \ \int_W^{W_0} G(W, V, h) dW. \tag{4.19}$$

Finally, if these equations are evaluated at the final weight, expressions for the *cruise distance* and *cruise time* result, that is,

$$x_f - x_0 \ = \ \int_{W_f}^{W_0} F(W, V, h) dW \tag{4.20}$$

$$t_f - t_0 \ = \ \int_{W_f}^{W_0} G(W, V, h) dW. \tag{4.21}$$

For each velocity profile $V(W)$, there exists a distance and a time. Once a velocity profile has been selected, the distance and the time for a given amount of fuel can be obtained. In general, the weight interval $W_0 - W_f$ is divided into n subintervals, and the integrals are rewritten as

$$
\begin{aligned}
x_f - x_0 \ &= \ \sum_{k=1}^n \int_{W_k}^{W_{k+1}} F(W, V, h) dW \\
t_f - t_0 \ &= \ \sum_{k=1}^n \int_{W_k}^{W_{k+1}} G(W, V, h) dW
\end{aligned}
\tag{4.22}
$$

where

$$W_1 = W_f, \quad W_{n+1} = W_0. \tag{4.23}$$

Then, the distance factor and the time factor are assumed to vary linearly with the weight over each subinterval, that is,

$$
\begin{aligned}
F \ &= \ F_k + \tfrac{F_{k+1}-F_k}{W_{k+1}-W_k}(W - W_k) \\
G \ &= \ G_k + \tfrac{G_{k+1}-G_k}{W_{k+1}-W_k}(W - W_k)
\end{aligned}
\tag{4.24}
$$

so that Eqs. (4.22) can be integrated analytically to obtain

$$x_f - x_0 \ = \ \sum_{k=1}^n \frac{1}{2}(F_{k+1} + F_k)(W_{k+1} - W_k) \tag{4.25}$$

$$t_f - t_0 \ = \ \sum_{k=1}^n \frac{1}{2}(G_{k+1} + G_k)(W_{k+1} - W_k). \tag{4.26}$$

In general, the number of intervals which must be used to get a reasonably accurate solution is small, sometimes just one.

While the distance and time can be computed for different velocity profiles such as constant velocity or constant lift coefficient, it is important for design purposes to find the maximum distance trajectory and the maximum time trajectory.

4.8   Cruise Point Performance for the SBJ

The analysis of the distance factor F and the time factor G is called *point performance* because only points of a trajectory are considered. For a fixed altitude, this can be done by plotting F and G versus velocity for several values of the weight. Regardless of the weight interval $[W_0, W_f]$ that is being used to compute distance and time, F and G can be computed for all values of W at which the airplane might operate. The use of F and G to compute the distance and time for a given velocity profile $V(W)$ and a given weight interval $[W_0, W_f]$ is called *path performance*, because a whole path is being investigated.

Point performance for the SBJ begins with the solution of Eq. (4.14) for the power setting $P(h, V, W)$ using Newton's method. It is shown in Fig. 4.3 for $h = 35.000$ ft. Note that the power setting is around 0.90. Then, P is substituted into Eqs. (4.17) to get the distance factor $F(W, V, h)$ and the time factor $G(W, V, h)$. Values of F and G have been computed for many values of the velocity (1 ft/s intervals) and for several values of the weight. These quantities are plotted in Figs. 4.4 and 4.5.

It is observed from Fig. 4.4 that the distance factor has a maximum with respect to the velocity for each value of the weight. This maximum has been found from the data used to compute F. At each weight, the velocity that gives the highest value of F is assumed to represent the maximum. Values of $V(W)$, $F_{max}(W)$, and the corresponding values of $G(W)$ are listed in Table 4.1 and plotted in Figs. 4.6 and 4.7. Note that V and $F_{max}(W)$ are nearly linear in W.

It is observed from Fig. 4.5 that the time factor has a maximum with respect to the velocity for each value of the weight. This maximum has been found from the data used to compute G. At each weight, the velocity that gives the highest value of GF is assumed to represent the

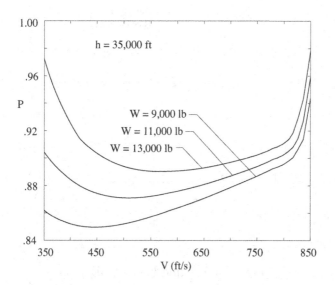

Figure 4.3: Power Setting (SBJ)

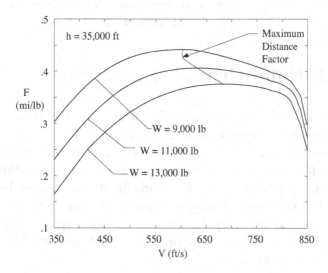

Figure 4.4: Distance Factor (SBJ)

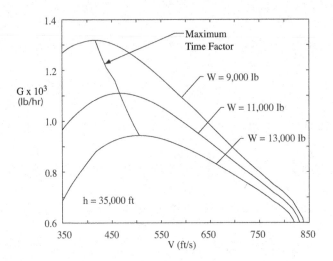

Figure 4.5: Time Factor (SBJ)

maximum. Values of $V(W)$, $G_{max}(W)$, and the corresponding values of $F(W)$ are listed in Table 4.1 and plotted in Figs. 4.6 and 4.7. Note that V and $G_{max}(W)$ are nearly linear in W.

Note that the velocity for maximum distance factor is roughly 35 % higher than the velocity for maximum time factor.

4.9   Optimal Cruise Trajectories

At this point, there are two approaches which can be followed. One is to specify a velocity profile, say for example $V=$ Const, and compute the distance and time for a given W_0 and W_f. The other is to find the velocity profile $V(W)$ that optimizes the distance or that optimizes the time. Because distance factor has a maximum, the optimal distance trajectory is a maximum. Similarly, because the time factor has a maximum, the optimal time trajectory is a maximum.. Optimal trajectories are considered first because they provide a yardstick with which other trajectories can be measured.

Table 4.1 SBJ Optimal Cruise Point Performance

$$h = 35,000 \text{ ft}$$

W	V	F_{max}	G	V	G_{max}	F
		Maximum Distance			Maximum Time	
lb	ft/s	mi/lb	hr/lb	ft/s	hr/lb	mi/lb
9,000	604	.442	1.073E-3	416	1.386E-3	.374
9,500	602	.434	1.058E-3	427	1.262E-3	.367
10,000	605	.424	1.031E-3	439	1.210E-3	.362
10,500	618	.415	.985E-3	456	1.161E-3	.361
11,000	631	.406	.943E-3	466	1.110E-3	.353
11,500	645	.398	.905E-3	477	1.064E-3	.345
12,000	658	.390	.870E-3	487	1.021E-3	.338
12,500	671	.383	.837E-3	496	.981E-3	.332
13,000	683	.376	.807E-3	506	.944E-3	.326

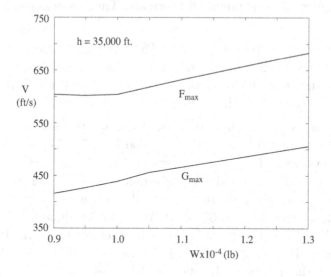

Figure 4.6: Optimal Velocity Profiles (SBJ)

4.9.1 Maximum distance cruise

The velocity profile $V(W)$ for maximizing the distance (4.20) is obtained
by maximizing the distance factor with respect to the velocity for each

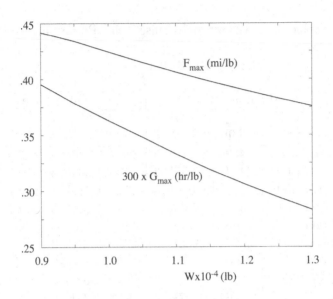

Figure 4.7: Maximum Distance and Time Factors (SBJ)

value of the weight (see Sec. 4.3). This velocity profile is then used to compute the maximum distance factor which is used to compute the maximum distance and to compute the time factor which is used to compute the time along the maximum distance trajectory.

This process has been carried out for the SBJ at h=35,000 ft with $W_0 = 12,000$ lb and $W_f = 10,000$ lb ($W_0 - W_f = 2,000$ lb of fuel). The values of F_{max} and G shown in Table 4.1 are used to compute the maximum distance and the corresponding time from Eqs. (4.25) and (4.26). As an example, to compute the maximum distance and the corresponding time for $h = 35,000$ ft, $W_0 = 12,000$ lb, and $W_f =10,000$ lb (2000 lb of fuel) using 5,000 lb weight intervals ($n = 4$), the values to be used in Eqs. (4.25) and (4.26) are listed in Table 4.2. Then, Eq. (4.25) gives the maximum distance of 813 mi (see Table 4.3). Similarly, the time along the maximum distance trajectory is obtained from Eq. (4.26) as 1.90 hr. This computation has also been made for one interval ($n = 1$), and the results agree well with those of $n = 4$. This happens because F_{max} (Fig. 4.5) and G are nearly linear in W.

Table 4.2 Maximum Distance Cruise

$h = 35,000$ ft, $W_0 = 12,000$ lb, $W_f = 10,000$ lb

k	W_k (lb)	$F_{max,k}$ (mi/lb)	G_k (hr/lb)
1	10,000	.424	.001031
2	10,500	.415	.000985
3	11,000	.406	.000943
4	11,500	.398	.000905
5	12,000	.390	.000870

Table 4.3: SBJ Optimal Cruise Path Performance

$h = 35,000$ ft, $W_0 = 12,000$ lb, $W_f = 10,000$ lb

		4 Intervals	1 Interval
Maximum	Distance (mi)	813	814
Distance	Time (hr)	1.90	1.90
Maximum	Distance (mi)	704	700
Time	Time (hr)	2.20	2.23

To actually fly the maximum distance velocity profile, it is necessary to know the weight as a function of time. If it is not available, the optimal path can only be approximated. Other velocity profiles are possible: constant lift coefficient, constant velocity, constant power setting, etc. The importance of the optimal profile is that the usefulness of the other profiles can be evaluated. For example, if a particular velocity profile is easy to fly and it gives a distance within a few percent of the maximum distance, it could be used instead. It can be shown that the maximum distance is almost independent of the velocity profile $V(W)$.

4.9.2   Maximum time cruise

The velocity profile $V(W)$ for maximizing the time (4.21) is obtained by maximizing the time factor with respect to the velocity for each value of

the weight (see Sec. 4.3). This velocity profile is then used to compute
the maximum time factor which is used to compute the maximum time
and to compute the distance factor which is used to compute the distance
along the maximum time trajectory.

This process has been carried out for the SBJ with $h = 35,000$
ft, $W_0 = 12,000$ lb, and $W_f = 10,000$ lb ($W_0 - W_f = 2,000$ lb of fuel). The
optimal velocity profile $V(W)$, the maximum time factor $G_{max}(W)$, and
the distance factor $F(W)$ are shown in Table 4.1. Then, the maximum
time and the corresponding distance are obtained from Eqs. (4.25) and
(4.26). For the maximum time of the SBJ, the maximum time has
been found by using four intervals to be 2.20 hr and the distance is 704
mi (Table 4.3). These results have also been obtained using only one
interval (700 mi and 2.23 hr). The agreement between the results using
one interval and the results obtained by using four intervals is very good.

4.10   Constant Velocity Cruise

In this section an example of arbitrarily specifying the velocity profile
is presented. The maximum distance path requires that the velocity
change as the weight changes (Table 4.1). Since there is no weight meter
on an airplane, the pilot cannot fly this trajectory very well. At constant
altitude, the indicated airspeed is proportional to the airspeed. Hence,
the pilot can fly a constant velocity trajectory fairly well even though
the controls must be adjusted to maintain constant velocity.

To obtain the distance for a particular velocity, the values of
the distance factor F for that velocity for several values of the weight are
used with Eq. (4.25). Similarly, the time along a constant velocity path
is obtained by using the values of the time factor G for that velocity for
several values of the weight and by using Eq. (4.26).

This process has been carried out for the SBJ operating at h
$= 35,000$ ft from $W_0 = 12,000$ lb to $W_f = 10,000$ lb (2,000 lb of fuel).
The results are shown in Table 4.4 for several values of the velocity
used for the cruise. With regard to the distance, note that there is a
cruise velocity for which the distance has a best value. This speed can
be calculated by computer or by curve fitting a parabola to the three
points containing the best distance. The curve fit leads to $V = 634$ ft/s
and $x_f - x_0 = 812$ mi. This value of the distance is almost the same

as the maximum distance $x_f - x_0 = 813$ mi. Hence, the airplane can be flown at constant velocity and not lose much distance relative to the maximum. It is emphasized that this conclusion could not have been reached without having the maximum distance path. As an aside, it is probably true that the airplane can be flown with any velocity profile (for example, constant power setting) and get close to the maximum distance.

A similar analysis with similar results can be carried out for the time.

Note that the term maximum distance is applied to the case where all possible velocity profiles are in contention for the maximum. On the other hand the term best distance is used for the case where the class of paths in contention for the maximum is restricted, that is, constant velocity paths. The maximum distance should be better than or at most equal to the best distance.

Table 4.4 Constant Velocity Cruise

$h = 35,000$ ft, $W_0 = 12,000$ lb, $W_f = 10,000$ lb

V (ft/s)	Distance (mi)	Time (hr)	V (ft/s)	Distance (mi)	Time (hr)
350	461	1.93	600	809	1.98
400	582	2.13	650	812	1.83
450	680	2.22	700	801	1.68
500	750	2.20	750	781	1.53
550	791	2.11	800	749	1.37

4.11 Quasi-steady Climb

The equations of motion for quasi-steady climbing flight are given by Eqs. (2.29), that is,

$$\dot{x} = V \tag{4.27}$$

$$\dot{h} = V\gamma \tag{4.28}$$

$$0 = T(h, V, P) - D(h, V, L) - W\gamma \tag{4.29}$$

$$0 = L - W \tag{4.30}$$

$$\dot{W} = -C(h, V, P)T(h, V, P) \tag{4.31}$$

This system of five equations has seven variables, $x(t), h(t), W(t), V(t),$ $\gamma(t), P(t),$ and $L(t)$. Hence, it has two mathematical degrees of freedom. Since it is easy to solve for L and γ, the degrees of freedom are associated with V and P. Experience shows that it is best to climb at maximum continuous thrust so the power setting is held constant leaving one degree of freedom, the velocity.

During the climb, an aircraft consumes around 5% of its weight in fuel. Hence, it is possible to assume that the weight of the aircraft is constant on the right-hand sides of the equations of motion. Then, the integration of the weight equation gives an estimate of the fuel consumed during the climb.

Since $L = W$, Eq. (4.28) can be solved for the flight path inclination or *climb angle* as

$$\gamma = \frac{T(h, V, P) - D(h, V, W)}{W} . \tag{4.32}$$

Two other important quantities are the *rate of climb*

$$\dot{h} = V\gamma = \left[\frac{T(h, V, P) - D(h, V, W)}{W}\right] V \tag{4.33}$$

and the *fuel factor*

$$H = -\frac{dh}{dW} = -\frac{\dot{h}}{\dot{W}} = \frac{\frac{[T(h,V,P)-D(h,V,W)]V}{W}}{C(h, V, P)T(h, V, P)}. \tag{4.34}$$

In order to be able to solve for the distance, the time, and the fuel in climbing from one altitude to another, the altitude is made the variable of integration. The differential equations of motion become the following:

$$\frac{dx}{dh} = \frac{1}{\gamma(h,V,P,W)}$$

$$\frac{dt}{dh} = \frac{1}{\dot{h}(h,V,P,W)} \tag{4.35}$$

$$-\frac{dW}{dh} = \frac{1}{H(h,V,P,W)},$$

where all variables are now functions of h. Since P and W are constant and $V = V(h)$, the integration can be performed to obtain $x(h), t(h)$, and $W(h)$:

$$x - x_0 = \int_{h_0}^{h} \frac{1}{\gamma(h, V, P, W)} \, dh \qquad (4.36)$$

$$t - t_0 = \int_{h_0}^{h} \frac{1}{\dot{h}(h, V, P, W)} \, dh \qquad (4.37)$$

$$W_0 - W = \int_{h_0}^{h} \frac{1}{H(h, V, P, W)} \, dh \ . \qquad (4.38)$$

Finally, if these equations are evaluated at the final altitude, the following expressions result for the distance, the time, and the fuel :

$$x_f - x_0 = \int_{h_0}^{h_f} \frac{1}{\gamma(h, V, P, W)} \, dh \qquad (4.39)$$

$$t_f - t_0 = \int_{h_0}^{h_f} \frac{1}{\dot{h}(h, V, P, W)} \, dh \qquad (4.40)$$

$$W_0 - W_f = \int_{h_0}^{h_f} \frac{1}{H(h, V, P, W)} \, dh. \qquad (4.41)$$

Hence, there are three possible optimal trajectories: minimum distance, minimum time, or minimum fuel.

Once the velocity profile is known, the distance, the time, and the fuel can be obtained by approximate integration. Here, the altitude interval h_0, h_f is divided into n subintervals, that is,

$$
\begin{aligned}
x_f - x_0 &= \sum_{k=1}^{n} \int_{h_k}^{h_{k+1}} dh/\gamma \\
t_f - t_0 &= \sum_{k=1}^{n} \int_{h_k}^{h_{k+1}} dh/\dot{h} \\
W_0 - W_f &= \sum_{k=1}^{n} \int_{h_k}^{h_{k+1}} dh/H
\end{aligned}
\qquad (4.42)
$$

where $h_1 = h_0$ and $h_{n+1} = h_f$. Next, it is assumed that the climb angle, the rate of climb, and the fuel factor vary linearly with the altitude over each altitude interval as follows:

$$
\begin{aligned}
\gamma &= \gamma_k + (\Delta\gamma/\Delta h)(h - h_k) \\
\dot{h} &= \dot{h}_k + (\Delta\dot{h}/\Delta h)(h - h_k) \\
H &= H_k + (\Delta H/\Delta h)(h - h_k)
\end{aligned}
\qquad (4.43)
$$

where

$$\Delta(\ ) = (\ )_{k+1} - (\ )_k. \tag{4.44}$$

If Eqs. (4.43) are substituted into Eqs. (4.42) and the integrations are performed, the following approximate expressions are obtained for the distance, the time, and the fuel consumed during the climb:

$$
\begin{aligned}
x_f - x_0 &= \sum_{k=1}^{n}(\Delta h/\Delta\gamma)\,\ln(\gamma_{k+1}/\gamma_k) \\
t_f - t_0 &= \sum_{k=1}^{n}(\Delta h/\Delta\dot h)\,\ln(\dot h_{k+1}/\dot h_k) \\
W_0 - W_f &= \sum_{k=1}^{n}(\Delta h/\Delta H)\,\ln(H_{k+1}/H_k)\ .
\end{aligned}
\tag{4.45}
$$

4.12   Climb Point Performance for the SBJ

Climb point performance involves the study of the flight path angle γ, the rate of climb $\dot h$ and the fuel factor H as defined in Eqs. (4.32) through (4.34). Values of these quantities have been computed for many values of the velocity (1 ft/s intervals) and several values of the altitude for $W = 11{,}000$ lb and $P = 0.98$.

The flight path angle is presented in Fig. 4.8. Note that the maximum γ occurs at sea level and is around 22 deg. As the altitude increases the maximum γ reduces to zero at the ceiling. At each altitude, the maximum γ is determined by finding the velocity (computed at 1 ft/s intervals) that gives the highest value of γ . This value of the velocity is used to compute the rate of climb and the fuel factor. Then, $V(h)$, $\gamma_{max}(h)$, $\dot h(h)$, and $H(h)$ are listed for several values of h in Table 4.5 and plotted in Figs. 4.11 and 4.12.

The rate of climb is shown in Fig. 4.9. Note that the maximum $\dot h$ occurs at sea level and is around 150 ft/s (9,000 ft/min). As the altitude increases the maximum $\dot h$ reduces to zero at the ceiling. At each altitude, the maximum $\dot h$ is determined by finding the velocity that gives the highest value of $\dot h$ which is computed at 1 ft/s intervals. This value of the velocity is used to compute the flight path angle and the fuel factor. Then, $V(h)$, $\gamma(h)$, $\dot h_{max}(h)$, and $H(h)$ are listed for several values of h in Table 4.6 and plotted in Figs. 4.11 and 4.12.

Because it is not possible to fly at the airplane ceiling, several other ceilings have been defined. The *service ceiling* is the altitude at which the maximum rate of climb is 100 ft/min. The *cruise ceiling* is

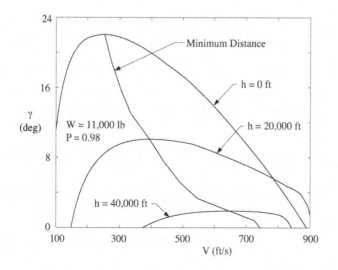

Figure 4.8: Climb Angle (SBJ)

Table 4.5 SBJ Optimal Climb Point Performance: γ

W=11,000 lb, P=.98

| | Maximum γ | | | |
| h | V | γ_{max} | $\dot{h}$ | H |
ft	ft/s	deg	ft/s	ft/lb
0	252	22.0	96.9	63.4
5,000	275	18.8	90.3	66.6
10,000	303	15.8	83.7	70.0
15,000	334	13.1	76.2	72.7
20,000	394	10.1	69.4	79.0
25,000	442	7.40	57.1	79.5
30,000	490	5.17	44.2	75.7
35,000	546	3.34	31.8	66.7
40,000	647	1.90	21.5	52.6
45,000	726	0.82	10.4	29.7

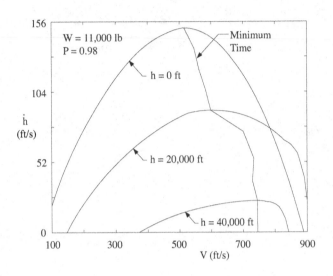

Figure 4.9: Rate of Climb (SBJ)

Table 4.6 SBJ Optimal Climb Point Performance: $\dot{h}$

W=11,000 lb, P=.98

h	\multicolumn{4}{c}{Maximum $\dot{h}$}			
h	V	γ	$\dot{h}_{max}$	H
ft	ft/s	deg	ft/s	ft/lb
0	515	16.8	151.	96.8
5,000	547	14.2	136.	96.5
10,000	562	12.3	120.	96.0
15,000	583	10.4	105.	94.7
20,000	599	8.63	90.2	92.4
25,000	701	6.09	74.5	84.9
30,000	734	4.27	54.7	75.9
35,000	730	2.95	37.6	66.5
40,000	745	1.79	23.3	51.6
45,000	745	0.81	10.9	20.8

the altitude where the maximum rate of climb is 300 ft/min. The *combat ceiling* is the altitude where the maximum rate of climb is 500 ft/min.

The fuel factor is presented in Fig. 4.10. Note that the maximum H occurs at sea level and is around 100 ft/lb. As the altitude increases the maximum H reduces to zero at the ceiling. At each altitude, the maximum H is determined by finding the velocity (computed at 1 ft/s intervals) that gives the highest value of γ. This value of the velocity is used to compute the rate of climb and the fuel factor. Then, $V(h)$, $\gamma(h)$, $\dot{h}(h)$, and $H_{max}(h)$ are listed for several values of h in Table 4.7 and plotted in Figs. 4.11 and 4.12.

4.13 Optimal Climb Trajectories

There are three possible performance indices for computing optimal climb trajectories: distance, time, or fuel. Because γ, $\dot{h}$, and H each have a maximum, the distance, time and fuel trajectories each have a minimum.

4.13.1 Minimum distance climb

The velocity profile for minimizing the distance (4.39) is obtained by maximizing the climb angle with respect to the velocity at each value of the altitude (Sec. 4.3). This velocity profile is used to compute the maximum climb angle, the rate of climb, and the fuel factor which are used to compute the minimum distance and the time and fuel along the minimum distance trajectory. This process has been carried out for the SBJ climbing from sea level to $h = 35,000$ ft with $W = 11,000$ lb and $P = 0.98$. The optimal velocity profile $V(h)$, the maximum climb angle $\gamma(h)$, the rate of climb $\dot{h}(h)$ and the fuel factor $H(h)$ shown in Table 4.5 are used to compute the minimum distance and the time and fuel along the minimum distance trajectory from Eqs. (4.45). These quantities have been found by using four intervals to be 42.2 mi, 9.22 min, and 484. lb respectively (Table 4.8). Note that the fuel is around 4% of the climb weight, justifying the approximation of weight constant on the right-hand sides of the climb equations of motion.

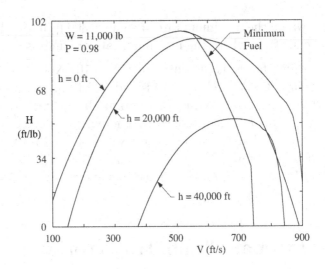

Figure 4.10: Fuel Factor (SBJ)

Table 4.7 SBJ Optimal Climb Point Performance: H

W=11,000 lb, P=.98

	Maximum H			
h	V	γ	$\dot{h}$	H_{max}
ft	ft/s	deg	ft/s	ft/lb
0	512	16.9	151.	96.8.
5,000	529	14.7	136.	96.7.
10,000	539	12.8	120.	96.3
15,000	553	10.9	105.	95.1
20,000	565	9.10	89.7	93.0
25,000	583	6.89	70.1	88.4
30,000	614	4.86	52.1	80.8
35,000	636	3.24	35.9	69.7
40,000	684	1.89	22.5	53.1
45,000	737	0.82	10.5	29.8

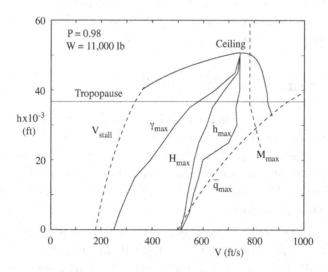

Figure 4.11: Optimal Velocity Profiles (SBJ)

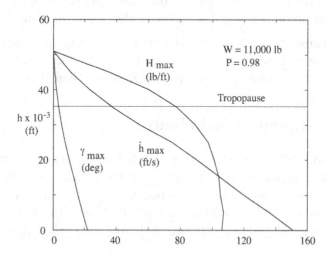

Figure 4.12: Maximum Climb Angle and Rate of Climb (SBJ)

These results have also been obtained using only one interval. The agreement between $n = 1$ and $n = 4$ is very good in γ and $\dot{h}$ because they are nearly linear in h. The error in H is around 10%.

4.13.2   Minimum time climb

The velocity profile for minimizing the time (4.40) is obtained by maximizing the rate of climb with respect to the velocity at each value of the altitude (Sec. 4.3). This velocity profile is used to compute the climb angle, the maximum rate of climb, and the fuel factor which are used to compute the minimum time and the distance and fuel along the minimum time trajectory.

This process has been carried out for the SBJ climbing from sea level to $h = 35,000$ ft with $W = 11,000$ lb and $P = 0.98$. The optimal velocity profile $V(h)$, the climb angle $\gamma(h)$, the maximum rate of climb $\dot{h}(h)$ and the fuel factor $H(h)$ shown in Table 4.6 are used to compute the distance, the time, and the fuel along the minimum time trajectory from Eqs. (4.45). These quantities have been found by using four intervals to be 51.4 mi, 6.97 min, and 399 lb respectively (Table 4.8). Note that the fuel is less than 4% of the climb weight, justifying the approximation of weight constant on the right-hand sides of the climb equations of motion.

These results have also been obtained using only one interval. The agreement between $n = 1$ and $n = 4$ is very good in γ and $\dot{h}$ because they are nearly linear in h. The error in H is less than 10%.

4.13.3   Minimum fuel climb

The velocity profile for minimizing the distance (4.41) is obtained by maximizing the fuel factor with respect to the velocity at each value of the altitude (Sec. 4.3). This velocity profile is used to compute the climb angle, the rate of climb, and the fuel factor which are used to compute the minimum fuel and the distance and time along the minimum fuel trajectory.

This process has been carried out for the SBJ climbing from sea level to $h = 35,000$ ft with $W = 11,000$ lb and $P = 0.98$. The optimal velocity profile $V(h)$, the climb angle $\gamma(h)$, the rate of climb $\dot{h}(h)$ and the maximum fuel factor $H(h)$ shown in Table 4.7 are used to compute the distance, the time, and the fuel along the minimum fuel trajectory from Eqs. (4.45). These quantities have been found by using four intervals to be 47.2 mi, 7.17 min, and 390 lb respectively (Table 4.8). Note that the fuel is less than 4% of the climb weight, justifying the approximation of weight constant on the right-hand sides of the climb equations of motion.

These results have also been obtained using only one interval. The agreement between $n = 1$ and $n = 4$ is very good in γ and $\dot{h}$ because they are nearly linear in h. The error in H is less than 10%.

Table 4.8 SBJ Optimal Climb Path Performance

$W = 11,000$ lb, $P = .98$, $h_0 = 0$ ft, $h_f = 35,000$ ft

		7 Intervals	1 Interval
	Distance (mi)	42.2	38.4
Minimum Distance	Time (min)	9.22	10.0
	Fuel (lb)	484.	538.
	Distance (mi)	51.4	47.7
Minimum Time	Time (min)	6.97	7.15
	Fuel (lb)	399.	433
	Distance (mi)	47.2	45.8
Minimum Fuel	Time (min)	7.17	7.28
	Fuel (lb)	390.	424.

4.14 Constant Equivalent Airspeed Climb

An example of selecting an arbitrary velocity profile is to assume that the airplane is flown at constant equivalent airspeed. Note that in Fig. 4.11 the optimal velocity profiles for the SBJ are functions of the altitude and may be difficult to fly. On the other hand, the pilot has an instrument for equivalent (indicated) airspeed, so it is possible to fly a constant equivalent airspeed trajectory. Here, the velocity profile is given by

$$V(h) = \frac{V_e}{\sqrt{\sigma(h)}} \qquad (4.46)$$

4.15   Descending Flight

In the mission profile, the descent segment is replaced by extending the cruise segment and, hence, is not very important. However, descending flight is just climbing flight for the case where the thrust is less than the drag. Here, the descent angle becomes negative so the *descent angle* is defined as $\phi = -\gamma$; the rate of climb becomes negative so the *rate of descent* is defined as $\dot{z} = -\dot{h}$; and the fuel factor is defined as dh/dW since both dh and dW are negative. The calculation of the minimum descent angle, the minimum descent rate, and the minimum fuel is the same as that for climbing flight, as is the calculation of the distance, time, and fuel.

Problems

All of the numbers in this chapter have been computed for the SBJ in App. A. Make similar computations for the airplane of Fig. 4.13 which is the SBJ of App. A with a lengthened fuselage to accommodate more passengers and with two Garrett TFE 731-2 turbofan engines. The take-off gross weight is 17,000 lb which includes 800 lb of reserve fuel and 6,200 lb of climb/cruise fuel.

1. Create functions that calculate the atmospheric properties, the drag, and the thrust and SFC. To calculate the drag, you need $C_{D_0}(M)$ and $K(M)$.

2. Calculate the flight envelope.

3. Calculate the maximum distance trajectory in cruise.

4. Calculate the minimum time trajectory in climb.

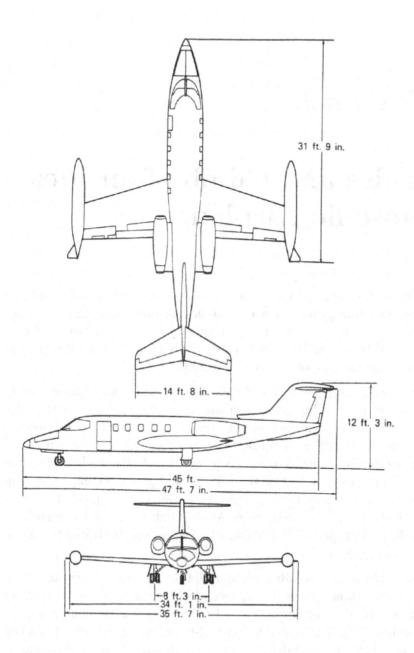

Figure 4.13: Turbofan Business Jet

Chapter 5

Cruise and Climb of an Ideal Subsonic Airplane

In Chap. 4, cruise and climb have been discussed for an arbitrary airplane. The data giving the aerodynamic and propulsion characteristics of these airplanes is given in the form of tables of numbers (subroutines), but they have been represented by functions of several variables. Trajectories must be obtained numerically.

The purpose of this chapter is to use an approximate analytical model for a subsonic jet airplane to derive analytical results for cruise and climb performance. The model is called the Ideal Subsonic Airplane (ISA) and is composed of a parabolic drag polar with constant coefficients, thrust independent of the velocity, and specific fuel consumption independent of the velocity and power setting. The drag polar is motivated by the fact that the optimal flight speeds for cruise and climb occur at speeds where Mach number effects are negligible. The thrust and specific fuel consumption forms are motivated by engine performance charts.

Analytical results are important for a variety of reasons. They expose important design parameters which might not be obvious from numerical results. They can be used to check extensive numerical computations. They can be used for back-of-the-envelope calculations. They can be used in iterative design codes to reduce the amount of computation.

At this point the reader should return to Chap. 4 and read the first few sections. The subjects covered were quasi-steady flight, mathematical degrees of freedom, special flight speeds, flight limitations, and trajectory optimization. Actually, this chapter is a continuation of Chap. 4, in that the functional relations $D(h, V, L), T(h, V, P)$, and $C(h, V, P)$ are replaced by formulas representing the aerodynamics and propulsion of the Ideal Subsonic Airplane. Then, analytical results are derived for the flight envelope, the distance and time in cruise, and the distance, time, and fuel in climb.

5.1 Ideal Subsonic Airplane (ISA)

The *Ideal Subsonic Airplane* (Sec. 3.9) has a parabolic drag polar with constant C_{D_0} and K, a thrust independent of the velocity, $T(h, P)$, and a specific fuel consumption independent of the velocity and power setting, $C(h)$. For this airplane the drag $D(h, V, L)$ is given by

$$D = \frac{1}{2}C_{D_0}\rho S V^2 + \frac{2KL^2}{\rho S V^2}. \tag{5.1}$$

where $\rho(h)$ is the density at the altitude the airplane is flying, S is the wing planform area, V is the velocity of the airplane, and L is the lift. The thrust and the specific fuel consumption are approximated by

$$T = T_t(P)(\rho/\rho_t)^a, \quad C = C_t(\rho/\rho_t)^b \tag{5.2}$$

where the subscript t denotes a value at the tropopause. These formulas are exact in the stratosphere where $a = 1$, and $b = 0$.

For quasi-steady flight, it is known that $L = W$. Hence, if h and W are given, the drag (5.1) has a minimum with respect to the velocity when

$$V = V^* \tag{5.3}$$

where

$$V^* = \sqrt{\frac{2W}{\rho S}}\sqrt{\frac{K}{C_{D_0}}}. \tag{5.4}$$

The minimum drag then has the value

$$D^* = 2\sqrt{C_{D_0}K}\, W. \tag{5.5}$$

From $L = W$, it is seen that the lift coefficient for minimum drag can be written as

$$C_L^* = \frac{2W}{\rho S V^{*2}}. \tag{5.6}$$

Hence, using Eq. (5.4) shows that the lift coefficient at the minimum drag condition is given by

$$C_L^* = \sqrt{\frac{C_{D_0}}{K}}. \tag{5.7}$$

The lift to drag ratio or aerodynamic efficiency is defined as

$$E = \frac{L}{D}. \tag{5.8}$$

For $L = W$ and h, W given, the lift to drag ratio has a maximum when the drag is a minimum so that

$$E^* = \frac{W}{D^*}. \tag{5.9}$$

Then, by using Eq. (5.5), it is seen that the maximum lift to drag ratio is given by

$$E^* = \frac{1}{2\sqrt{C_{D_0}K}}. \tag{5.10}$$

In deriving the performance formulas, it is possible to use either C_{D_0}, K or C_L^*, E^* since either set can be derived from the other. The latter is used because the resulting formulas have a simpler form.

Analytical manipulations can be simplified by using the *nondimensional speed*

$$u = \frac{V}{V^*}, \quad V^* = \sqrt{\frac{2W}{\rho S C_L^*}}. \tag{5.11}$$

In terms of u, Eq. (5.1) for the drag becomes

$$D = \frac{W}{2E^*}\left(u^2 + \frac{n^2}{u^2}\right), \tag{5.12}$$

where $n = L/W$ is the load factor. For $L = W$ ($n=1$), the expression for the drag becomes

$$D = \frac{W}{2E^*}\left(u^2 + \frac{1}{u^2}\right). \tag{5.13}$$

If h, W are given, V^* is given, and u is proportional to V. Hence, minimization of the drag with respect to V and minimization with respect to u are the same operation. Note that the drag has a minimum with respect to u when $u=1$, Eq. (5.3), and has the minimum value $D^* = W/E^*$, Eq. (5.9).

The SBJ is the Subsonic Business Jet of App. A which is powered by two GE CJ610-6 turbojet engines. The ISBJ is the Ideal Subsonic Business Jet whose aerodynamic constants are given by $C_{D_0} = .023$ and $K = .073$ and whose engine characteristics T_t and C_t are given in Table 3.7 with $a = 1.2$, $b = 0.1$ in the troposphere and $a = 1$, $b = 0$ in the stratosphere.

To see how the formulas are used, consider the Ideal Subsonic Business Jet (ISBJ) at h=35,000 ft, W=11,000 lb, and P=.98. The following numbers are given, can be looked up, or can be calculated:

$$
\begin{aligned}
&h = 35,000 \text{ ft}, \quad \rho = .000737 \text{ slug/ft}^3 \\
&a = 973 \text{ ft/s}, \quad \rho_t = .000706 \text{ slug/ft}^3 \\
&C_{D_0} = .023, \quad K = .073, \quad C_L^* = .561, \quad E^* = 12.2 \\
&W = 11,000 \text{ lb}, \quad S = 232 \text{ ft}^2 \\
&V^* = 480 \text{ ft/s}, \quad D^* = 902 \text{ lb} \\
&P = 0.98, \quad T_t = 1420 \text{ lb}, \quad a = 1.2, \quad T = 1490 \text{ lb} \\
&C_t = 1.18 \text{ 1/hr}, \quad b = 0.1, \quad C = 1.19 \text{ 1/hr} = .000329 \text{ 1/s} \\
&V_{stall} = 322 \text{ ft/s}, \quad M_{max} = 0.81, \quad V_{M_{max}} = 788 \text{ ft/s}
\end{aligned}
\tag{5.14}
$$

Note that all computations must be performed in the ft, lb, s, rad system even though results may be given in mi, min, hr, or deg.

5.2 Flight Envelope

As stated in Sec. 4.5, the flight envelope is the region of the altitude-velocity plane where an airplane can maintain steady level flight. For the ISA, Eq. (4.9) becomes

$$
T(h, P) - \frac{W}{2E^*}\left(u^2 + \frac{1}{u^2}\right) = 0.
\tag{5.15}
$$

Since the minimum drag ($u = 1$) is $D^* = W/E^*$, the minimum thrust required is $T_{min} = D^*$. Hence, a *nondimensional thrust* τ is defined as

$$
\tau = \frac{T}{T_{min}} = \frac{T}{D^*} = \frac{T}{W/E^*} \geq 1.
\tag{5.16}
$$

As a consequence, Eq. (5.15) becomes

$$\tau - \frac{1}{2}\left(u^2 + \frac{1}{u^2}\right) = 0 \tag{5.17}$$

and can be rewritten as

$$u^4 - 2\tau u^2 + 1 = 0. \tag{5.18}$$

For a given altitude, weight and power setting (given τ), this quadratic equation in u^2 can be solved for the velocity as

$$u = \sqrt{\tau \pm \sqrt{\tau^2 - 1}}. \tag{5.19}$$

The minus sign gives the low-speed solution, and the plus sign gives the high-speed solution.

Consider flight of the ISBJ at $h = 35,000$ ft, $W = 11,000$ ft, and P=0.98 for which the data of Eq. (5.14) is valid. Let the low-speed solution be denoted by subscript 1, and the high-speed solution, by subscript 2. It is seen that τ=1.66, $u_1 = .580$, and $u_2 =1.73$. Also, with $V^* = 480$ ft/s, $V_1 = 278$ ft/s and $V_2 = 828$ ft/s. Since $V_{stall} = 322$ ft/s, the low-speed solution is below the stall speed and does not exist. It occurs at a lift coefficient where the parabolic drag polar is not valid. Also, the high-speed solution is above $V_{M_{max}}$, meaning that the airplane is not allowed to operate at the maximum speed at this altitude.

By varying the altitude, the curve in the altitude-velocity plane defining the level flight speeds can be generated. Combined with whatever speed restrictions are imposed on the aircraft, the region below this curve is called the *flight envelope*. See for example Fig. 4.2. For the ISBJ the $T - D = 0$ curve is about 4,000 ft lower than that of Fig. 4.2. The difference is caused by assuming thrust independent of the velocity.

The *ceiling* occurs when there is only one solution for the velocity, that is, when

$$\tau = 1 \quad \text{and} \quad u = 1. \tag{5.20}$$

Use of the analytical expression (5.2) for T leads to following value of the density at the ceiling:

$$\rho = \rho_t [W/(T_t E^*)]^{1/a} . \tag{5.21}$$

For most jet aircraft, the ceiling occurs in the stratosphere (a=1). For the ISBJ, Eq. (5.14), the density and altitude of the ceiling are given by

ρ= 4.48 E-4 slugs/ft^3 and h=45,500 ft. The altitude is computed from the formulas of the standard atmosphere for the constant temperature part of the stratosphere. Finally, the velocity associated with the ceiling is given by V=614 ft/s.

5.3 Quasi-steady Cruise

Next to be discussed are the distance and time during a constant altitude cruise. These quantities are also called the *range* and *endurance*. However, because there is considerable distance associated with the climb, the range might be taken to be the sum of the distance in climb and the distance in cruise.

The equations for the *cruise distance* and the *cruise time* are given by Eqs. (4.20) and (4.21) where the distance factor and the time factor for the ISA are given by

$$\begin{aligned} F(W,V,h) &= \tfrac{V}{CT} = \tfrac{V}{C(h)D(h,V,W)} \\ G(W,V,h) &= \tfrac{1}{CT} = \tfrac{1}{C(h)D(h,V,W)} \end{aligned} \tag{5.22}$$

Since $C = C(h)$, it is not necessary to solve for the power setting in that the thrust can be replaced by the drag $(T = D)$. Then, for the ISA, Eq. (5.13) leads to

$$\begin{aligned} F(W,V,h) &= \frac{V^*}{C} \frac{u}{\frac{W}{2E^*}\left(u^2+\frac{1}{u^2}\right)} \\ G(W,V,h) &= \frac{1}{C} \frac{1}{\frac{W}{2E^*}\left(u^2+\frac{1}{u^2}\right)} \end{aligned} \tag{5.23}$$

where $u = V/V^*$. Finally,

$$x_f - x_0 = \frac{2E^*}{C} \int_{W_f}^{W_0} \frac{V^*}{W} \frac{u^3}{u^4+1}\, dW \tag{5.24}$$

$$t_f - t_0 = \frac{2E^*}{C} \int_{W_f}^{W_0} \frac{1}{W} \frac{u^2}{u^4+1}\, dW \tag{5.25}$$

These equations give the distance and the time of the ISA in terms of the unknown velocity profile $V(W)$. At this point, the distance and the time can be integrated for a particular velocity or for the optimal velocity profile. The optimal velocity profiles are derived first so that the results for particular velocity profiles can be evaluated.

5.4   Optimal Cruise Trajectories

Since the distance factor has a maximum (see for example Fig. 4.4) with respect to the velocity, there is a maximum distance trajectory. Similarly, because the time factor has a maximum (see for example Fig. 4.5) with respect to the velocity, there is a maximum time trajectory.

5.4.1   Maximum distance cruise

Since the altitude is constant in cruise and the weight is constant during optimization, u is proportional to V. Hence, the velocity for maximum distance is obtained by maximizing F with respect to u. It is given by $u = \sqrt[4]{3}$ or equivalently

$$V = \sqrt[4]{3}V^* = \sqrt[4]{3}\sqrt{\frac{2W}{\rho S C_L^*}}.\tag{5.26}$$

For the ISBJ, Eq. (5.14), the speed for maximum distance varies between 700 and 550 ft/s as the weight decreases from 13,000 to 9,000 lb.

Next, the distance and time integrals become

$$
\begin{aligned}
x_f - x_0 &= \tfrac{3^{3/4}}{2}\tfrac{E^*}{C}\left[\tfrac{2}{\rho S C_L^*}\right]^{1/2}\int_{W_f}^{W_0}\frac{1}{W^{1/2}}\,dW \\[2mm]
t_f - t_0 &= \tfrac{3^{1/2}}{2}\tfrac{E^*}{C}\int_{W_f}^{W_0}\frac{1}{W}\,dW
\end{aligned}
\tag{5.27}
$$

and can be integrated to obtain the maximum distance and the time along the maximum distance path:

$$
\begin{aligned}
x_f - x_0 &= 3^{3/4}\tfrac{E^*}{C}\sqrt{\tfrac{2W_0}{\rho S C_L^*}}\left[1-\sqrt{\tfrac{W_f}{W_0}}\right] \\[2mm]
t_f - t_0 &= \tfrac{3^{1/2}}{2}\tfrac{E^*}{C}\ln\left[\tfrac{W_0}{W_f}\right].
\end{aligned}
\tag{5.28}
$$

Consider the cruise of the ISBJ at $h = 35{,}000$ ft with $W_0 = 12{,}000$ lb and $W_f = 10{,}000$ lb ($W_{fuel} = 2000$ lb). With the data of Eq. (5.14), these formulas yield a maximum distance of 698 mi and a flight time of 1.63 hr.

The expression for the maximum distance can be solved for the minimum fuel required for a given distance as follows:

$$\frac{W_{fuel}}{W_0} = 1 - \left(1 - \frac{C(x_f - x_0)}{3^{3/4} E^*} \sqrt{\frac{\rho S C_L^*}{2W_0}}\right)^2. \tag{5.29}$$

Note that the fuel is a function of both the distance and the weight at which the cruise is begun. There are two comments that can be made about this formula. First, it can be shown that the fuel decreases as W_0 decreases. Second, in the stratosphere where C is constant, the formula says that the fuel decreases with the altitude. However, as the cruise altitude increases, more fuel is required to get to that altitude. It would seem that there is an optimal climb-cruise trajectory for maximum range. Another way to define the cruise altitude is to require that the maximum rate of climb be a particular value.

5.4.2 Maximum time cruise

The velocity profile for maximum time is obtained by differentiating the time factor (5.23) with respect to u. Maximum time occurs when $u = 1$ or equivalently

$$V = V^*. \tag{5.30}$$

For the ISBJ, Eq. (5.14), the optimal velocity varies between 500 and 430 ft/s as the weight decreases from 13,000 and 9,000 lb.

With this result, Eqs. (5.24) and (5.25) become

$$
\begin{aligned}
t_f - t_0 &= \frac{E^*}{C} \int_{W_f}^{W_0} \frac{1}{W} \, dW \\
x_f - x_0 &= \frac{E^*}{C} \left[\frac{2}{\rho S C_L^*}\right]^{1/2} \int_{W_f}^{W_0} \frac{1}{W^{1/2}} \, dW.
\end{aligned}
\tag{5.31}
$$

Then, the maximum time and the distance along the maximum time path are given by

$$
\begin{aligned}
t_f - t_0 &= \frac{E^*}{C} \ln \left[\frac{W_0}{W_f}\right] \\
x_f - x_0 &= \frac{2E^*}{C} \sqrt{\frac{2W_0}{\rho S C_L^*}} \left[1 - \sqrt{\frac{W_f}{W_0}}\right]
\end{aligned}
\tag{5.32}
$$

Consider the cruise of the ISBJ at $h = 35{,}000$ ft with $W_0 = 12{,}000$ lb and $W_f = 10{,}000$ lb ($W_{fuel} = 2000$ lb). With the data of

Eq. (5.14), these formulas yield a maximum time of 1.88 hr and a corresponding distance of 618 mi.

Note that the maximum distance (698 mi) is greater the distance along the maximum time path (618 mi) as it should be. Similarly, the maximum time (1.88 hr) is greater than the time along the maximum distance path (1.63 hr).

5.4.3   Remarks

From a design point of view, the aerodynamic configuration of an aircraft being designed for maximum distance should be made such that the quantity $E^*/\sqrt{C_L^*}$ is as high as possible. In terms of C_{D_0} and K, this means that $1/(2C_{D_0}^{3/4}K^{1/4})$ should be as high as possible. On the other hand, an airplane being designed for maximum time should have an E^* or $1/(2C_{D_0}^{1/2}K^{1/2})$ as high as possible. With regard to the engines, they should be designed for low specific fuel consumption.

While only optimal velocity profiles have been presented so far, it is possible to fly other velocity profiles and lose only a small amount of distance. Cruise at constant velocity is discussed in the next section. Cruise at constant lift coefficient is considered in Prob. 5.1; and cruise at constant power setting is discussed in Prob. 5.2. The importance of the optimal velocity profile is that the loss incurred by flying some other velocity profile can be assessed.

5.5   Constant Velocity Cruise

Because the optimal velocity (distance or time) varies with the weight, it is not easily flown by a pilot. On the other hand, the airspeed indicator at constant altitude is proportional to the velocity so that cruise at constant velocity can easily be flown by a pilot. A question to be answered is how much distance or time is lost by flying this velocity profile instead of the optimum.

The purpose of this section is to give an example of computing the distance for a given velocity profile, in this case, constant velocity. Here,

$$u = \frac{V}{V^*} = V\sqrt{\frac{\rho SC_L^*}{2W}} \qquad (5.33)$$

which contains the weight. Substitution into the distance equation (5.24) leads to

$$x_f - x_0 = \frac{V}{C}\int_{W_0}^{W_f} \frac{dW}{\frac{1}{2}C_{D_0}\rho SV^2 + \frac{2KW^2}{\rho SV^2}} \qquad (5.34)$$

as it should. This integral has the form

$$\int \frac{dx}{a^2 + b^2x^2} = \frac{1}{ab}\tan^{-1}\frac{bx}{a} \qquad (5.35)$$

where

$$a = \sqrt{\frac{C_{D_0}\rho SV^2}{2}}, \quad b = \sqrt{\frac{2K}{\rho SV^2}}. \qquad (5.36)$$

Hence, the distance becomes

$$x_f - x_0 = \frac{V}{C}\frac{1}{ab}\left[\tan^{-1}\frac{bW_0}{a} - \tan^{-1}\frac{bW_f}{a}\right]. \qquad (5.37)$$

Because of the identity

$$\tan^{-1}x - \tan^{-1}y = \tan^{-1}\frac{x-y}{1+xy}, \qquad (5.38)$$

Eq. (5.37) becomes after some rearranging

$$x_f - x_0 = \frac{V}{C}\frac{1}{ab}\tan^{-1}\frac{\frac{bW_0}{a}\left(1 - \frac{W_f}{W_0}\right)}{1 + (\frac{bW_0}{a})^2\frac{W_f}{W_0}}. \qquad (5.39)$$

After substitution of a and b, the following formula for the distance is obtained for a constant velocity cruise from W_0 to W_f:

$$x_f - x_0 = \frac{2E^*V}{C}\tan^{-1}\frac{\frac{2W_0}{C_L^*\rho SV^2}\left(1 - \frac{W_f}{W_0}\right)}{1 + (\frac{2W_0}{C_L^*\rho SV^2})^2\frac{W_f}{W_0}}. \qquad (5.40)$$

The ratio W_f/W_0 is related to the fuel fraction W_{fuel}/W_0 as

$$\frac{W_f}{W_0} = 1 - \frac{W_{fuel}}{W_0}. \qquad (5.41)$$

Note that the distance depends on both the amount of fuel consumed and the airplane weight at the beginning of the cruise.

It is possible to solve for the minimum fuel required to go a given distance as

$$\frac{W_{fuel}}{W_0} = 1 - \frac{\frac{2W_0}{C_L^*\rho SV^2} - \tan\frac{C(x_f-x_0)}{2E^*V}}{\frac{2W_0}{C_L^*\rho SV^2}\left(1+\frac{2W_0}{C_L^*\rho SV^2}\tan\frac{C(x_f-x_0)}{2E^*V}\right)}. \tag{5.42}$$

In Sec. 5.4.1, the velocity profile which maximizes the distance is found by considering all possible velocity profiles. Hence, that velocity profile is the maximum. In this section, only constant velocity profiles are considered. It is possible to find the best constant velocity, but it is not in general as good as the maximum.

At what speed should the airplane be flown to get the best constant-speed distance? This question can be answered by taking a derivative of Eq. (5.40) with respect to V. The resulting equation must be solved numerically. The other possibility is to calculate the distance (5.40) for a lot of values of the velocity and take the highest value. Either way, for the data (5.14), the best distance for the ISBJ occurs at the velocity $V = 630$ ft/s and has the value 697 mi. Note that the maximum distance is given by 698 mi. Hence, flying at the best constant velocity gives a distance which is close to the optimum and can be used with confidence because the optimum is known.

Note that the time along a constant velocity path is given by

$$t_f - t_0 = \frac{x_f - x_0}{V}. \tag{5.43}$$

The question of which speed should be flown for the best constant-speed time is answered in the same manner as that for the best constant-speed distance.

5.6   Quasi-steady Climb

The distance, time, and fuel for a quasi-steady climb from h_0 to h_f are given by Eqs. (4.39) through (4.41), that is,

$$x_f - x_0 = \int_{h_0}^{h_f} \frac{1}{\gamma} \, dh \tag{5.44}$$

$$t_f - t_0 = \int_{h_0}^{h_f} \frac{1}{\dot{h}} \, dh \qquad (5.45)$$

$$W_0 - W_f = \int_{h_0}^{h_f} \frac{1}{H} \, dh. \qquad (5.46)$$

In these equations, γ is the flight path angle, $\dot{h}$ is the rate of climb, and H is the fuel factor given by Eqs. (4.32) through (4.34). For the Ideal Subsonic Airplane, γ, $\dot{h}$, and H can be written in terms of the nondimensional velocity u and the nondimensional thrust τ as

$$\gamma = \frac{1}{E^*}\left[\tau - \frac{1}{2}\left(u^2 + \frac{1}{u^2}\right)\right] \qquad (5.47)$$

$$\dot{h} = \frac{V^*}{E^*}\left[\tau u - \frac{1}{2}\left(u^3 + \frac{1}{u}\right)\right] \qquad (5.48)$$

$$H = \frac{V^*}{WC_\tau}\left[\tau u - \frac{1}{2}\left(u^3 + \frac{1}{u}\right)\right]. \qquad (5.49)$$

These equations give the distance, time, and fuel written in terms of the unknown velocity profile $V(h)$. Hence, given the velocity profile, these equations can be integrated, in principle, to give formulas for the distance, time, and fuel.

5.7   Optimal Climb Trajectories

Because the drag has a minimum with respect to the velocity, the climb angle, the rate of climb, and the fuel factor all have a maximum with respect to the velocity. Hence, there are three possible optimal trajectories: (1) minimum distance, (2) minimum time, and (3) minimum fuel. Each optimal trajectory is found by minimizing one of the integrals in Eqs. (5.44) through (5.46) with respect to the velocity. Since the weight is fixed and the altitude is fixed while differentiating with respect to the velocity, V^* is constant, and u is directly proportional to V. Hence, optimal climb trajectories are obtained by minimizing with respect to u.

It is not possible to obtain analytical solutions for the distance, time and fuel of the ISBJ along each of the optimal paths. Their values can be computed using Eqs. (4.45). The numbers for doing so are listed in Table 5.1. Note that the maximum flight path angle is 27.0 deg and the maximum rate of climb is 204 ft/s (12,000 ft/min). They both occur at sea level where the thrust is the highest.

5.7.1   Minimum distance climb

From Eqs. (5.44) and (5.47), it is seen that the *minimum distance climb* is obtained by maximizing the climb angle γ with respect to u at each altitude. The optimal velocity is given by $u=1$ or

$$V = V^* \tag{5.50}$$

and the maximum flight path angle is

$$\gamma = \frac{\tau - 1}{E^*}. \tag{5.51}$$

The rate of climb and the fuel factor along this path are given by

$$\dot{h} = \frac{V^*(\tau-1)}{E^*}$$
$$H = \frac{V^*(\tau-1)}{WC\tau}. \tag{5.52}$$

The minimum distance climb has been computed for the ISBJ. The numbers needed to do so are shown in Table 5.1, and the distance, time, and fuel are given in Table 5.2. Note that the minimum distance is 41.8 mi, and the corresponding time and fuel are 9.59 min and 571 lb.

Table 5.1 ISBJ Optimal Climb Point Performance

W=11,000 lb, P=.98

h	τ	Minimum Distance				Minimum Time			
		V	γ_{max}	$\dot{h}$	H	V	γ	$\dot{h}_{max}$	H
ft		ft/s	deg	ft/s	ft/lb	ft/s	deg	ft/s	ft/lb
0	6.76	268	27.0	126.	55.8	572	20.5	204.	90.4
5,000	5.65	287	21.8	1.09	58.9	564	16.9	166.	89.4
10,000	4.70	310	17.4	94.0	61.8	558	13.7	134.	88.0
15,000	3.88	336	13.5	79.2	64.2	553	11.0	106.	85.8
20,000	3.18	365	10.2	65.1	65.5	550	8.55	82.1	82.5
25,000	2.58	398	7.42	51.6	64.9	548	6.42	61.5	77.4
30,000	2.08	436	5.06	38.5	61.3	550	4.53	43.6	69.4
35,000	1.66	479	3.08	25.8	52.4	557	2.87	27.9	56.7
40,000	1.31	537	1.43	13.4	35.5	578	1.38	14.0	36.9
45,000	1.03	606	0.12	1.31	4.52	610	0.12	1.32	4.54

Table 5.2 ISBJ Optimal Climb Path Performance

$W = 11,000$ lb, $P = .98$, $h_0 = 0$ ft, $h_f = 35,000$ ft, $\Delta h = 5,000$ ft

Minimum Distance	Distance (mi)	41.8
	Time (min)	9.59
	Fuel (lb)	571
Minimum Time	Distance (mi)	48.7
	Time (min)	7.74
	Fuel (lb)	439

5.7.2 Minimum time climb

The *minimum time climb* is obtained by maximizing the rate of climb (5.48) at each altitude. The optimal velocity profile is given by

$$V = \frac{V^*}{\sqrt{3}} \sqrt{\tau + \sqrt{\tau^2 + 3}}, \tag{5.53}$$

and the maximum rate of climb has the value

$$\dot{h} = \frac{2V^*}{3^{3/2}E^*}\sqrt{\tau + \sqrt{\tau^2 + 3}}\,[2\tau - \sqrt{\tau^2 + 3}\,]. \tag{5.54}$$

Next, the flight path angle and the fuel factor along this path become

$$\begin{aligned}
\gamma &= \tfrac{2}{3E^*}[2\tau - \sqrt{\tau^2 + 3}\,] \\
H &= \tfrac{2V^{*2}}{WCE^*_\tau}\sqrt{\tau + \sqrt{\tau^2 + 3}}\,[2\tau - \sqrt{\tau^2 + 3}\,].
\end{aligned} \tag{5.55}$$

The minimum time climb has been computed for the ISBJ. The numbers needed to do so are shown in Table 5.1, and the distance, time, and fuel are given in Table 5.2. Note that the minimum time is 7.74 min, and the corresponding distance and fuel are 48.7 min and 439 lb.

5.7.3   Minimum fuel climb

The fuel factor has a maximum with respect to the velocity. However, because $C = C(h)$ and $T = T(h, P)$, the *minimum fuel climb* is identical with the minimum time climb. Another way of saying this is that the velocity profile for minimum fuel is the same as that for maximum rate of climb.

5.8   Climb at Constant Equivalent Airspeed

Note that the optimal velocity profiles for the distance and the time are not easy for a pilot to fly because they depend on the altitude. On the other hand, a climb at constant equivalent airspeed is easily flown because the pilot can fly at constant indicated airspeed which is approximately the equivalent airspeed.

As an example of prescribing a velocity profile, consider a climb of the ISA at constant equivalent airspeed (same as constant dynamic pressure). For such a climb, $V_e = \sqrt{\sigma}V$ so that the true airspeed is

$$V = \frac{V_e}{\sqrt{\sigma}} \tag{5.56}$$

which is the velocity profile $V(h)$. This means that

$$u = \frac{V}{V^*} = \frac{V_e}{\sqrt{\sigma}V^*} = \frac{V_e}{V_s^*} \tag{5.57}$$

where V_s^* is the velocity for maximum lift-to-drag ratio at sea level. The result is that $u = $ Const along this climb.

For a particular value of u, the quantities $\gamma(h)$, $\dot{h}(h)$, and $H(h)$ are given by Eqs. (5.47) through (5.49) and can be integrated for the distance, time and fuel using Eqs. (5.44) through (5.46). Analytical results cannot be obtained for a standard atmosphere, because τ is a complicated function of altitude. For the turbojet, analytical results can be obtained for an exponential atmosphere for distance and time but not for fuel. Since the numerical procedure must be used for the fuel, it might as well be used for distance and time as well.

For the turbofan, it is possible to use $T = T_t(P)(\rho/\rho_t)$ and $C = C_t =$ Const throughout the atmosphere. For this case, the exponential atmosphere $\rho = \rho_0 \exp(-h/\lambda)$ gives analytical results for the distance, time and fuel. Here, for $u = $ Const,

$$x_f - x_0 = -\frac{E^*\lambda}{d}\left[\eta + \ln(\tau_f e^{-\eta} - d)\right]_{\eta_0}^{\eta_f} \tag{5.58}$$

$$t_f - t_0 = \frac{E^*\lambda}{V_e}\sqrt{\frac{\sigma_t}{\tau_t d}}\left[\ln\frac{\sqrt{\tau_t e^{-\eta}} + \sqrt{d}}{\sqrt{\tau_t e^{-\eta}} + \sqrt{d}}\right]_{\eta_0}^{\eta_f} \tag{5.59}$$

$$W_0 - W_f = \frac{2WC}{V_e\sqrt{\tau_t}}\left[-\sqrt{e^{-\eta}} + \frac{1}{2}\sqrt{\frac{d}{\tau_t}}\ln\frac{\sqrt{\tau_t e^{-\eta}} + \sqrt{d}}{\sqrt{\tau_t e^{-\eta}} + \sqrt{d}}\right]_{\eta_0}^{\eta_f} \tag{5.60}$$

where

$$d = \frac{1}{2}\left(u^2 + \frac{1}{u^2}\right), \quad \eta = \frac{h - h_t}{\lambda}. \tag{5.61}$$

The final question is what value of V_e should be flown for best distance, for best time, and for best fuel. How do these values compare with the optimal values?

5.9 Descending Flight

Descending flight occurs when $T < D$. The analysis is the same as for climbing flight with two exceptions. First, the minimum distance path becomes a maximum distance path. Second, the minimum time path becomes a maximum time path.

Problems

5.1 Consider the ISA in a constant altitude cruise. If the airplane is also flown at constant lift coefficient, show that the velocity profile $V(W)$ is given by

$$V(W) = \sqrt{\frac{2W}{\rho S C_L}}.$$

Next, show that the distance and the time are

$$x_f - x_0 = \frac{2E(C_L)}{C}\sqrt{\frac{2}{\rho SC_L}}(\sqrt{W_0} - \sqrt{W_f})$$

$$t_f - t_0 = \frac{E(C_L)}{C}\ln\frac{W_0}{W_f}.$$

With regard to the airframe, the distance has its highest value when C_L is chosen to maximize $E/\sqrt{C_L}$ where

$$E = \frac{C_L}{C_{D_0} + KC_L^2}.$$

What is this C_L, and what is the highest value of $E/\sqrt{C_L}$? If the airplane is flown at the C_L for maximum L/D, what is the corresponding value of $E/\sqrt{C_L}$? Show that flying at E^* leads to a 12% reduction in the distance relative to the best distance.

5.2 Consider the ISA in a constant power setting (constant thrust) cruise. By solving the equation $T(h, P) - D(h, V, W) = 0$ for the velocity, show that the high speed solution for $V(W)$ is given by

$$V = \sqrt{\frac{2TE^*}{\rho SC_L^*}}\sqrt{1 + \sqrt{1 - \left(\frac{W}{TE^*}\right)^2}}.$$

Next, show that the distance integral becomes

$$x_f - x_0 = \frac{E^*}{C}\sqrt{\frac{2TE^*}{\rho SC_L^*}}\int_{\mu_f}^{\mu_0}\sqrt{1 + \sqrt{1 - \mu^2}}\,d\mu$$

where $\mu = W/TE^*$. By using the change of variables $z = \sqrt{1 - \mu^2}$, show that the distance integral can be integrated to give

$$x_f - x_0 = \frac{2E^*}{3C}\sqrt{\frac{2TE^*}{\rho SC_L^*}}[A(W_0) - A(W_f)]$$

where

$$A(W) = \left(2 + \sqrt{1 - \left(\frac{W}{TE^*}\right)^2}\right)\sqrt{1 - \sqrt{1 - \left(\frac{W}{TE^*}\right)^2}}.$$

Consider the distance for the ISBJ at h=35,000 ft when W_0 =12,000 lb and $W_f = 10,000$ lb. Plot the distance versus T to find the thrust for best distance and the best distance. Compare the result with the maximum distance of 698 mi. Conclusion?

5.3 Derive the equation for the time for Problem 5.2.

5.4 The equations of motion for gliding flight $(T = 0, W = $ Const$)$ of an arbitrary airplane are given by

$$\dot{x} = V$$

$$\dot{h} = V\gamma$$

$$0 = D(h, V, L) + W\gamma$$

$$0 = L - W.$$

How many mathematical degrees of freedom are there? Change the variable of integration to the altitude and recompute the number of mathematical degrees of freedom.

Show that the distance and time can be written as

$$x_f - x_0 = \int_{h_f}^{h_0} \frac{dh}{\phi(h, V, W)}$$

$$t_f - t_0 = \int_{h_f}^{h_0} \frac{dh}{\dot{z}(h, V, W)}$$

where the glide angle $\phi \triangleq -\gamma$ and the rate of descent $\dot{z} \triangleq -\dot{h}$ are given by

$$\phi = \frac{D(h, V, W)}{W}$$

$$\dot{z} = \frac{D(h, V, W)V}{W}.$$

Hence, maximum distance is achieved by flying at the speed for minimum glide angle at each altitude, and maximum time, by flying at the speed for minimum rate of descent at each altitude.

5.5 For the ISA, show that maximum distance in glide occurs when $u = 1$ and that the minimum glide angle is $\phi = 1/E^*$. Next, show that the maximum distance is

$$x_f - x_0 = E^*(h_0 - h_f)$$

and that the corresponding time is

$$t_f - t_0 = \frac{E^*}{V_s^*} \int_{h_f}^{h_0} \sqrt{\sigma} \, dh.$$

where V_s^* is V^* at sea level. Evaluate the integral for an exponential atmosphere.

Note that a glider being designed for maximum distance should have a high E^*.

5.6 For the Ideal Subsonic Airplane, show that maximum time in glide occurs when $u = 1/\sqrt[4]{3}$, that the minimum rate of descent is

$$\dot{z} = \frac{2\sqrt[4]{3}}{3} \frac{V^*}{E^*}$$

and the maximum time is

$$t_f - t_0 = \frac{3}{2\sqrt[4]{3}} \frac{E^*}{V_s^*} \int_{h_f}^{h_0} \sqrt{\sigma} \, dh.$$

Also, show that the distance is

$$x_f - x_0 = \frac{\sqrt{3}}{2} E^*(h_0 - h_f).$$

Aerodynamically, how should a glider be designed for maximum time?

5.7 For gliding flight of the ISA, show that the distance and time achieved along a constant lift coefficient flight path is given by

$$x_f - x_0 = E(C_L)(h_0 - h_f)$$

$$t_f - t_0 = \frac{E}{V_s^*} \int_{h_f}^{h_0} \sqrt{\sigma} \, dh.$$

At what C_L should the glider be flown to achieve the best distance? Show that a constant C_L path is the same as a constant equivalent airspeed path.

5.8 Show that the distance in gliding flight of the ISA along a constant velocity trajectory is given by

$$x_f - x_0 = \int_{h_f}^{h_0} \frac{W \, dh}{(1/2)C_{D_0}\rho SV^2 + 2KW^2/\rho SV^2}$$

which must be integrated numerically for the $\rho(h)$ from the standard atmosphere. Show that the integration can be performed if the exponential atmosphere is used and the density is made the variable of integration.

Note that the time is given by

$$t_f - t_0 = \frac{x_f - x_0}{V}$$

5.9 The climb-cruise allows the airplane to gain altitude as the airplane loses weight. It can be studied by assuming quasi-steady flight and $W\gamma << T$ or D. Consider the ISA in the stratosphere ($T = T_t\rho/\rho_t$, $C = C_t$), and assume that the airplane is being flown at constant power setting and constant lift coefficient. Show that the altitude varies with the weight as

$$\rho = \rho_t W/(T_t E)$$

and that the velocity is constant, that is,

$$V = \sqrt{2T_t E/(\rho_t S C_L)}.$$

Finally, show that the distance is given by

$$x_f - x_0 = \frac{E^{3/2}}{CC_L^{1/2}} \sqrt{\frac{2T_t}{\rho_t S}} \ln \frac{W_0}{W_f}$$

and that the C_L that maximizes the distance is

$$C_L = C_L^*/\sqrt{2}.$$

Compare the climb-cruise with the constant altitude cruise (Prob. 5.1).

Chapter 6

Take-off and Landing

In this chapter, the take-off and landing segments of an airplane trajectory are studied. After the segments are defined, high-lift devices are discussed, and a method for predicting their aerodynamics is given. Next, the equations of motion for the ground run are derived and solved for distance. Then, specific formulas are obtained for the take-off ground distance and the landing ground distance. Transitions from take-off to climb and descent to landing are investigated so that take-off distance and landing distance can be estimated.

6.1 Take-off and Landing Definitions

The purpose of this section is to define the take-off and landing maneuvers so that the take-off and landing distances can be determined. It is assumed that the altitude is constant during the take-off and landing ground runs and that the altitude is sea level.

Take-off

The take-off segment of an aircraft trajectory is shown in Fig. 6.1. The aircraft is accelerated at constant power setting and at a constant angle of attack (all wheels on the ground) from rest to the *rotation speed* V_R. For safety purposes, the rotation speed is required to be somewhat greater than the stall speed, and it is taken here to be

$$V_R = 1.2V_{stall} = 1.2\sqrt{\frac{2W}{\rho S C_{Lmax}}} \ . \tag{6.1}$$

When the rotation speed is reached, the aircraft is rotated over a short time to an angle of attack which enables it to leave the ground at the *lift-off speed* V_{LO} and begin to climb. The transition is also flown at constant angle of attack and power setting. The take-off segment ends when the aircraft reaches an altitude of $h = 35$ ft.

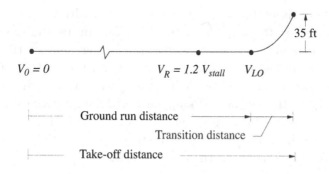

Figure 6.1: Take-off Definitions

Because airplanes are designed essentially for efficient cruise, they are designed aerodynamically for high lift-to-drag ratio. A trade-off is that the maximum lift coefficient decreases as the lift-to-drag ratio increases. This in turn increases the stall speed, increases the rotation speed, and increases the take-off distance. Keeping the take-off distance within the bounds of existing runway lengths is a prime consideration in selecting the size (maximum thrust) of the engines. The same problem occurs on landing but is addressed by using flaps. A low flap deflection can be used on take-off to reduce the take-off distance.

The take-off segment is composed of a *ground run* and a *transition* to climb. During the ground run portion, the drag polar of the airplane includes the drag of the landing gear, the drag and lift of the flaps, and a reduction in the induced drag due to the presence of the ground. The ground prevents the air moving over the wing from being deflected as far downward as it is in free flight. Hence, the resultant aerodynamic force is not rotated as far backward, thus decreasing the induced drag. In terms of the parabolic drag polar, the zero-lift drag coefficient is increased, and the induced drag factor is decreased. As an example, the SBJ might have $C_{D_0} = 0.023$, $K = 0.073$ in free flight, but it would have $C_{D_0} = 0.064$, $K = 0.060$ during the ground run with flaps set at 20 deg. During the transition, the ground effect decreases with altitude and landing gear and flap retraction is begun.

In the analysis which follows, the take-off is considered under the assumption that the rotation speed and the lift-off speed are identical. In other words, the aircraft is accelerated to the lift-off speed and then rotated instantaneously to the angle of attack for transition.

There is another element of take-off for multi-engined airplanes. This occurs when the airplane loses an engine during the ground run. When this happens, the pilot must decide whether to abort the take-off or to take-off on one less engine and fly around the pattern and land. This maneuver is analyzed by defining a speed, say V_D. If the engine fails before V_D, the airplane is stopped by braking. If the engine fails after V_D, the take-off is completed. For some V_D the take-off distance equals the accel/stop distance. This distance is called the *balanced field length*, and this V_D is called the *decision speed*.

Landing

The landing segment of an aircraft trajectory is shown in Fig. 6.2. Landing begins with the aircraft in a reduced power setting descent at an altitude of $h = 50$ ft with gear and flaps down. As the aircraft nears

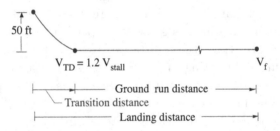

Figure 6.2: Landing Definitions

the ground, it is flared to rotate the velocity vector parallel to the ground. The aircraft touches down on the main gear and is rotated downward to put the nose gear on the ground. Then, brakes and sometimes reverse thrust, spoilers, and a drag chute are used to stop the airplane. The landing ends when the aircraft comes to rest. For safety purposes, the *touchdown speed* is required to be somewhat greater than the stall speed and is taken here to be

$$V_{TD} = 1.2V_{stall}. \tag{6.2}$$

The landing segment is composed of a *transition* and a *ground run*. During the transition, the airplane approaches the runway along the *glide slope* and at some point flares to rotate the velocity vector parallel to the ground. The airplane is in ground effect so that its lift increases and its induced drag decreases as the aircraft nears the ground. In addition the drag polar is affected by landing gear drag, flap drag, and flap lift as in take-off. However, flaps are at the highest setting to produce the highest $C_{L_{max}}$, the most drag, and the lowest touchdown speed. The ground run begins at touchdown. The airplane is rotated to the ground run attitude (all wheels on the ground); brakes are applied to increase the coefficient of rolling friction; thrust is reversed; and spoilers are extended. Thrust reversers deflect the jet stream of an engine so that the thrust acts opposite to the direction of motion. As much as 40% of the forward thrust can be achieved during reversal. Spoilers are metal panels located on the top of the wing forward of the flaps. When they are rotated into the air, they increase the drag, and they spoil the flow over the rest of the wing and reduce the lift, thereby increasing the reaction force of the runway on the airplane and, hence, the friction force. Finally, drag chutes are used when everything else fails to stop an aircraft in available runway lengths. Their effect is to increase the drag coefficient. For the SBJ, the parabolic drag polar during the ground run with the gear down and flaps down at 40 deg is given by $C_{D_0} = .083$ and $K = .052$.

To achieve an analytical solution. it is assumed that the aircraft touches down at $V_{TD} = 1.2 V_{stall}$ and rotates instantaneously to the ground roll attitude. Brakes are applied at touchdown. The ground run ends when the aircraft comes to rest.

6.2 High-lift Devices

The maximum lift coefficient plays an important role in determining the take-off and landing distances of an airplane. There are two types of devices which are used to increase the maximum lift coefficient: *slats* and *flaps*.

The slat is a *leading edge device* and is shown in Fig. 6.3. A complicated mechanism extends and rotates the slat to an effective position. The slat increases $C_{L_{max}}$ without changing C_{L_α} or α_{0L} as shown in Fig. 6.4.

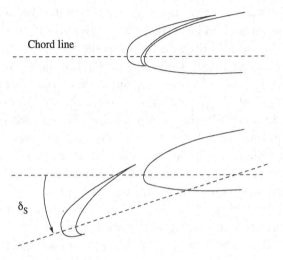

Figure 6.3: Closed and Open Slat

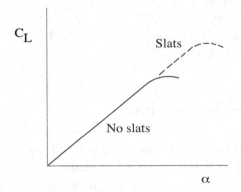

Figure 6.4: Effect of Slats

There are several types of *flaps* as shown in Fig. 6.5, where positive flap deflection is trailing edge down. In general, flaps make α_{0L} more negative, do not change C_{L_α}, and increase $C_{L_{max}}$ (see Fig, 6.6). The effect of a flap deflection on the drag polar is shown in Fig. 6.7. This polar differs from the regular polar in that the minimum drag point is above the axis. While a more general parabola can be fit to this polar (see Prob. 3.5), it is also possible to fit the regular parabola to it and realize that it is not accurate for both low and high lift coefficients.

6.3 Aerodynamics of High-Lift Devices

The lift coefficient and drag coefficient of an airplane in free flight are given in terms the angle of attack α by

$$
\begin{aligned}
C_L &= C_{L_\alpha}(\alpha - \alpha_{0L}) \\
C_D &= C_{D_0} + K C_{L_\alpha}^2 (\alpha - \alpha_{0L})^2
\end{aligned}
\tag{6.3}
$$

For take-off and landing C_{D_0} and K are taken to be the values in free flight at $M = 0.2$. If the angle of attack is the same, but the landing gear is down and the flaps are deflected, the force coefficients become

$$
\begin{aligned}
C_L &= C_{L_\alpha}(\alpha - \alpha_{0L}) + \Delta C_{L_F} \\
C_D &= C_{D_0} + \Delta C_{D_{lg}} + \Delta C_{D_F} + [K/f(\delta_F)]C_{L_\alpha}^2(\alpha - \alpha_{0L})^2
\end{aligned}
\tag{6.4}
$$

The quantity $\Delta C_{D_{lg}}$ is the increase in C_{D_0} due to the landing gear; it is estimated from the formula

$$
\Delta C_{D_{lg}} = .0032\, W_{TO}^{0.8}/S
\tag{6.5}
$$

where W_{TO} is the *take-off gross weight* and S is the wing planform area. Note that the flaps have three effects on C_L and C_D. The lift coefficient is increased by the amount ΔC_{L_F}; the zero-lift drag coefficient is increased by the amount ΔC_{D_F}; and the induced drag factor is increased by the factor $1/f(\delta_F)$, Fig. 6.8. The prediction of these quantities is discussed in the next section.

The drag coefficient can be written in terms of the lift coefficient so that C_L and the drag polar are given by

$$
\begin{aligned}
C_L &= C_{L_\alpha}[(\alpha - \alpha_{0L}) + \Delta C_{L_F}] \\
C_D &= C_{D_0} + \Delta C_{D_{lg}} + \Delta C_{D_F} + [K/f(\delta_F)][C_L - \Delta C_{L_F}]^2.
\end{aligned}
\tag{6.6}
$$

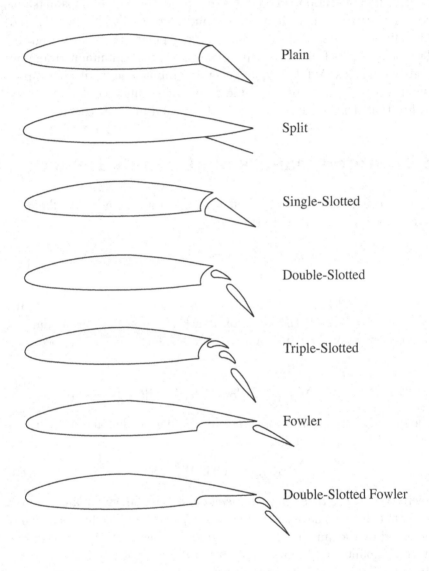

Plain

Split

Single-Slotted

Double-Slotted

Triple-Slotted

Fowler

Double-Slotted Fowler

Figure 6.5: Types of Flaps

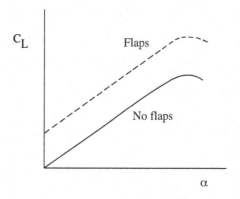

Figure 6.6: Effect of Flaps

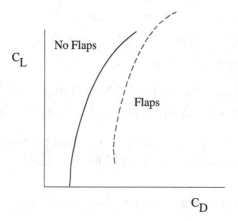

Figure 6.7: Effect of Flaps on the Drag Polar

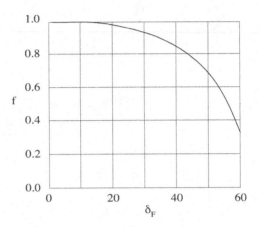

Figure 6.8: Effect of Flap Deflection on Induced Drag Factor

If the airplane is near the ground, an interesting phenomenon called *ground effect* occurs. In general, the pressure below the wing increases thereby increasing the lift. Also, the aerodynamic force is not rotated as far back so that the induced drag is decreased. The expressions for C_L and C_D become

$$
\begin{aligned}
C_L &= G_L(\bar{h})[C_{L_\alpha}(\alpha - \alpha_{0L}) + \Delta C_{L_F}(\delta_F)] \\
C_D &= C_{D_0} + \Delta C_{D_{lg}} + \Delta C_{D_F}(\delta_F) \\
&+ G_D(\bar{h})[K/f(\delta_F)][C_L - \Delta C_{L_F}(\delta_F)]^2
\end{aligned}
\tag{6.7}
$$

where $G_L \geq 1$ and $G_D \leq 1$ are given by

$$
\begin{aligned}
G_L &= 1.0 + (0.00211 - 0.0003(A_W - 3.0))e^{5.2(1-\bar{h}/b_W)} \\
G_D &= 1.111 + 5.55\bar{h}/b_W - [29.8(\bar{h}/b_W + 0.02)^2 + 0.817]^{1/2}.
\end{aligned}
\tag{6.8}
$$

In these relations, $\bar{h}$ is the height of the flap trailing edge above the ground. The equation for G_D is valid for $\bar{h} < 0.9b_W$, otherwise, $G_D = 1.0$.

The equation for the drag polar has the form

$$
C_D = C_{D_m} + K_m(C_L - C_{L_m})^2
\tag{6.9}
$$

where

$$
\begin{aligned}
C_{D_m} &= C_{D_0} + \Delta C_{D_{lg}} + \Delta C_{D_F}(\delta_F) \\
K_m &= G_D(\bar{h})[K/f(\delta_F)] \\
C_{L_m} &= \Delta C_{L_F}(\delta_F).
\end{aligned}
\tag{6.10}
$$

It is possible to fit this polar with the standard form $C_D = C_{D_0} + KC_L^2$ with the understanding that it does not fit the actual polar at low and high values of C_L.

6.4 ΔC_{L_F}, ΔC_{D_F}, and $C_{L_{max}}$

In this section formulas are provided for estimating ΔC_{L_F}, ΔC_{D_F}, and $C_{L_{max}}$. The method used to compute these values is based on the knowledge of the flap characteristics for the following reference conditions (Ref. Sc): $\lambda_W = 1$, $A_W = 12$, $(t/c)_W = 0.10$, $\Lambda_{qc_W} = 0$, $\delta_S = 45$ deg, and $\delta_F = 60$ deg. The reference values for the increment in lift coefficient due to the flaps $(\Delta C_{L_F})_r$, the increase of the lift coefficient due to slats $(\Delta C_{L_S})_r$, the increment in drag coefficient due to flaps $(\Delta C_{D_F})_r$, and the maximum lift coefficient $(C_{L_{max}})_r$ are shown in Table 6.1. These reference values are corrected for the actual wing geometry and the actual slat and flap deflections (δ_S and δ_F).

Table 6.1. Reference Values for Flap Input Variables

Flap Type	Plain	Split	Single-Slotted	Double-Slotted	Triple-Slotted	Fowler	Two-slot Fowler
$(\delta_F)_r$	60	60	60	60	60	60	60
$(\delta_S)_r$	45	45	45	45	45	45	45
$\Delta(C_{L_F})_r$	0.90	0.80	1.18	1.40	1.60	1.67	2.25
$\Delta(C_{L_S})_r$	0.93	0.93	0.93	0.93	0.93	0.93	0.93
$\Delta(C_{D_F})_r$	0.12	0.23	0.13	0.23	0.23	0.10	0.15
$(C_{L_{max}})_r$	1.4	1.4	1.4	1.4	1.4	1.4	1.4

The size of the slat is given by c_S/c_W, and it is assumed that the slats extend over the entire exposed leading-edge of the wing; thus,

$$b_S/b_W = 1 - d_B/b_W \qquad (6.11)$$

The size of the flap is given by c_F/c_W and b_F/b_W.

In terms of the correction factors K_i, the increment in lift coefficient due to flaps is given by

$$\Delta C_{L_F} = (\Delta C_{L_F})_r K_3 K_4 K_5 K_6 K_7 K_8 K_{13} K_{14}, \qquad (6.12)$$

and the increment in drag coefficient due to the flaps is given by

$$\Delta C_{D_F} = (\Delta C_{D_F})_r K_{16} K_{17} K_{18} K_{19} . K_{20} \tag{6.13}$$

The maximum lift coefficient for the airplane in the clean configuration is given by

$$C_{L_{max}} = K_{15} + K_{13} K_{14} K_1 K_2 (C_{L_{max}})_r \tag{6.14}$$

while the maximum lift coefficient with flaps and slats extended is

$$
\begin{aligned}
C_{L_{max}} = K_{15} + K_{13} K_{14} [(C_{L_{max}})_r K_1 K_2 \\
+ (\Delta C_{L_F})_r K_3 K_4 K_5 K_6 K_7 K_8 + (\Delta C_{L_S})_r K_9 K_{10} K_{11} K_{12}].
\end{aligned}
\tag{6.15}
$$

The correction factors K_i are obtained from the plots in Figs. 6.10 through 6.29. They are functions of geometry, flight condition, and deflection angles. These plots are located at the end of the chapter because there are so many of them. In Fig. 6.28 the sweep of the flap is given by

$$\Lambda_F = \Lambda_{te} + \tan^{-1}[4(c_F/c_W)/3(\tan\Lambda_{qc} - \tan\Lambda_{te})] \tag{6.16}$$

6.5   Ground Run

In Fig. 6.9, the forces acting on the airplane during the ground run are shown (Ref. Mi1). In addition to the usual thrust, drag, lift and weight forces, there are the *reaction force R* due to the runway and the *friction force f* due to the rolling of the wheels about their axles and on the runway. Actually, these forces act at the wheels but have been moved to the center of gravity and accompanied by moments which are not shown.

Since the altitude is constant, there is no motion in the z direction, and the flight path inclination is zero. Hence, the kinematic equation in the x direction, the dynamic equations along the tangent and the normal to the flight path, and the weight equation are written as

$$
\begin{aligned}
\dot{x} &= V \\
\dot{V} &= (g/W)[T\cos(\alpha + \varepsilon_0) - D - f] \\
0 &= T\sin(\alpha + \varepsilon_0) + L + R - W \\
\dot{W} &= -CT
\end{aligned}
\tag{6.17}
$$

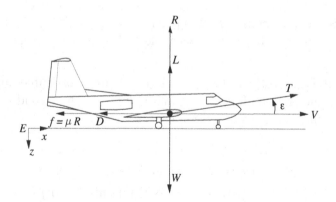

Figure 6.9: Forces Acting on the Airplane During the Ground Run

In general, thrust and specific fuel consumption obey function relations of the form $T = T(h, V, P)$ and $C = C(h, V, P)$. Here, however, the altitude is constant, and the power setting is held constant. It is shown later that the ground run distance is minimized by maximum power setting. For the ground run, these functional relations become

$$T = T(V) , \quad C = C(V) \qquad (6.18)$$

Next, the lift and drag coefficients obey functional relations of the form $C_L = C_L(\alpha, M, \delta_F)$ and $C_D = C_D(\alpha, M, R_e, \delta_F)$. Mach number effects are negligible because the speeds are low, and the Reynolds number is held constant. Since the angle of attack and flap setting are constant, the lift coefficient and the drag coefficient are constant along the ground run. Finally, since drag and lift are defined by

$$D = (1/2)C_D\rho S V^2, \quad L = (1/2)C_L\rho S V^2 \qquad (6.19)$$

and the altitude is constant, the drag and the lift satisfy functional relations of the form

$$D = D(V), \quad L = L(V) \qquad (6.20)$$

along the ground run.

Finally, the *coefficient of rolling friction* is defined as $\mu \triangleq f/R$ so that the friction force is modeled as

$$f = \mu R \qquad (6.21)$$

where μ is assumed constant. A typical value is $\mu = 0.02$ for take-off on a concrete runway. In landing with the brakes applied, μ increases to 0.3 - 0.4.

If Eqs. (6.18), (6.20), and (6.21) are substituted into Eqs. (6.17) and the reaction force is eliminated between the second and third equations, the equations of motion can be rewritten as

$$
\begin{aligned}
\dot{x} &= V \\
\dot{V} &= (g/W)\{T(V)\cos(\alpha + \varepsilon_0) - D(V) \\
&\quad -\mu[W - L(V) - T(V)\sin(\alpha + \varepsilon_0)]\} \\
\dot{W} &= -C(V)T(V).
\end{aligned}
\tag{6.22}
$$

This system of equations has zero mathematical degrees of freedom (three variables $x(t)$, $V(t)$, $W(t)$ and three equations) and can in principle be solved.

To obtain an analytical solution, it is assumed that $\alpha + \varepsilon_0$ is small and that $T(\alpha + \varepsilon_0) << W$. Also, since the weight of the fuel consumed on take-off (less than 50 lb for the SBJ) is negligible with respect to the initial weight, the weight is assumed to be constant on the right-hand side of the equations of motion during the ground run. This assumption causes the weight equation to uncouple from the system, and it can be solved once $V(t)$ is known to obtain the fuel consumed during take-off. While it is possible to solve the equations of motion with the thrust in the form

$$
T = P(h) + Q(h)V + R(h)V^2,
\tag{6.23}
$$

it is assumed that the thrust is constant, equal to an average value over the speed range encountered during the ground run. The specific fuel consumption is also assumed constant. These assumptions are consistent with the Ideal Subsonic Airplane assumptions.

Because of the above approximations. the equations of motion can be rewritten as

$$
\begin{aligned}
\dot{x} &= V \\
\dot{V} &= (g/W)[T - \mu W - (1/2)\rho S(C_D - \mu C_L)V^2] \\
\dot{W} &= -CT
\end{aligned}
\tag{6.24}
$$

Since the only variable on the right hand side of the equations of motion is the velocity, the velocity is chosen to be the variable of integration.

Hence, the equation for the ground run distance becomes

$$\frac{dx}{dV} = \frac{W}{g} \frac{V}{(T - \mu W) - (1/2)\rho S(C_D - \mu C_L)V^2}. \tag{6.25}$$

This equation can be integrated to obtain

$$x = -\frac{W}{g\rho S(C_D - \mu C_L)} \ln[A(V)] + \text{Const} \tag{6.26}$$

where

$$A(V) = (T - \mu W) - (1/2)\rho S(C_D - \mu C_L)V^2. \tag{6.27}$$

It holds for both the take-off and landing ground runs. The corresponding equations for the time and the fuel are easily obtained, and they are left as exercises.

6.5.1 Take-off ground run distance

In the analysis which follows, the take-off is considered under the assumption that the rotation speed and the lift-off speed are identical. It is assumed that the aircraft is accelerated from rest to the lift-off speed, rotated instantaneously to the angle of attack for transition, and leaves the ground. The value of μ for take-off is taken to be 0.02 for a concrete runway. From typical values for the SBJ, it is seen that $T - \mu W > 0$ and $C_D - \mu C_L > 0$.

The boundary conditions for the take-off ground run are as follows:

$$V_0 = 0, \quad x_0 = 0$$
$$V_f = V_{LO} = V_R = 1.2V_{stall}. \tag{6.28}$$

Application of the initial conditions in Eq. (6.26) leads to

$$\text{Const} = \frac{W}{g\rho S(C_D - \mu C_L)} \ln[A(0)] \tag{6.29}$$

and

$$x - x_0 = -\frac{W}{g\rho S(C_D - \mu C_L)} \ln\left[\frac{A(V)}{A(0)}\right] \tag{6.30}$$

which gives the distance as a function of the velocity. Then, the application of the final condition gives the *take-off ground run distance* as

$$x_f - x_0 = -\frac{W}{g\rho S(C_D - \mu C_L)} \ln\left[1 - \frac{\rho S(C_D - \mu C_L)V_{LO}^2}{2(T - \mu W)}\right] \tag{6.31}$$

where the minus sign is due to the logarithm being negative.

With respect to the power setting, it is easily seen that maximum power setting (maximum thrust) produces minimum ground run distance. The situation is not clear with respect to the flap setting. In general, $C_L = C_L(\alpha, \delta_F)$ and $C_D = \bar{C}_{D_0}(\delta_F) + \bar{K}(\delta_F)C_L^2$ where δ_F is the flap angle. Since C_D and C_L increase with flap setting and the lift-off or stall speed decreases, it is not possible to tell what the product $(C_D - \mu C_L)V_{LO}^2$ does without calculating some numbers.

6.5.2   Landing ground run distance

The equation for landing distance versus velocity is the same as that for the take-off ground run, Eq. (6.26). The airplane is assumed to touch down, rotate instantaneously to the ground roll attitude (all wheels on the ground), and apply brakes. The last causes μ to be in the range 0.3-0.4. The thrust at idle is sufficiently small or reverse thrust ($T < 0$) is applied so that $T - \mu W < 0$. Finally, even though the drag coefficient includes gear and flaps, $C_D - \mu C_L < 0$.

The boundary conditions for the landing ground run are given by

$$V_0 = V_{TD} = 1.2V_{stall}, \quad x_0 = 0$$
$$V_f = 0. \tag{6.32}$$

Application of the initial conditions leads to

$$\text{Const} = \frac{W}{g\rho S(C_D - \mu C_L)} \ln[A(V_{TD})] \tag{6.33}$$

so that the ground run distance versus velocity becomes

$$x - x_0 = -\frac{W}{g\rho S(C_D - \mu C_L)} \ln\left[\frac{A(V)}{A(V_{TD})}\right]. \tag{6.34}$$

Finally, applying the final condition leads to the following expression for the *landing ground run distance*:

$$x_f - x_0 = \frac{W}{g\rho S(C_D - \mu C_L)} \ln\left[1 - \frac{\rho S(C_D - \mu C_L)V_{TD}^2}{2(T - \mu W)}\right]. \tag{6.35}$$

The landing distance formula yields the effect of various parameters. For example, if reverse thrust ($T < 0$) is applied, it is seen that landing distance decreases as $-T$ increases. Spoilers increase C_D and decrease C_L; hence, they decrease x_f since $dx_f/d(C_D - \mu C_L)$ is negative. Also, a drag chute increases C_D and decreases x_f since $dx_f/dC_D < 0$. The effect of flaps must be determined by computation unless V_{TD} is fixed at some value.

6.6 Transition

The objective is to obtain an estimate for the distance traveled while the airplane climbs to $h = 35$ ft or descends from $h = 50$ ft. The analysis begins with the nonsteady equations of motion which have two mathematical degrees of freedom. It is assumed that the airplane is flown at constant angle of attack and constant power setting. There are now zero mathematical degrees of freedom so that the equations can be solved, in principle. All of the assumptions and approximations which are made are now summarized.

a. $\alpha = $ Const and $P = $ Const

b. Fixed geometry: landing gear down and flaps down

c. Negligible ground effect: $C_D = $ Const and $C_L = $ Const

d. Small altitude interval: $\rho = $ Const

e. $\gamma^2 << 1, \quad \varepsilon^2 << 1, \quad T\varepsilon << W$

f. V and W are constant on the right hand side of the equations of motion

With these assumptions the nonlinear equations of motion (Chap. 2) reduce to the following:

$$
\begin{aligned}
\dot{x} &= V \\
\dot{h} &= V\gamma \\
\dot{V} &= (g/W)(T - D - W\gamma) \\
\dot{\gamma} &= (g/V)(n - 1) \\
\dot{W} &= -CT
\end{aligned}
\qquad (6.36)
$$

where the load factor $n \triangleq L/W$ is constant. The velocity and weight
equations uncouple from the system and can be solved later to get velocity and weight changes. If the altitude is used as the variable of
integration, the remaining equations become

$$
\begin{aligned}
\frac{dx}{dh} &= \frac{1}{\gamma} \\
\frac{dt}{dh} &= \frac{1}{V\gamma} \\
\frac{d\gamma}{dh} &= \frac{g(n-1)}{V^2\gamma}
\end{aligned}
\tag{6.37}
$$

6.6.1   Take-off transition distance

The boundary conditions for the take-off transition at sea level are given
by

$$
\begin{aligned}
h_0 = 0, \quad x_0 = 0, \quad \gamma_0 = 0, \\
h_f = 35 \text{ ft.}
\end{aligned}
\tag{6.38}
$$

Then, integration of the γ equation leads to

$$
\gamma = \overset{+}{-} \sqrt{\frac{2g(n-1)h}{V^2} + 2C}.
\tag{6.39}
$$

For take-off, γ is positive, and application of the initial conditions gives

$$
\gamma = (1/V)\sqrt{2g(n-1)h}.
\tag{6.40}
$$

At the final point, it is seen that

$$
\gamma_f = (1/V)\sqrt{2g(n-1)h_f}.
\tag{6.41}
$$

Next, the distance equation can be integrated subject to the initial and
final conditions to obtain the *take-off transition distance*

$$
x_f - x_0 = V\sqrt{\frac{2h_f}{g(n-1)}}.
\tag{6.42}
$$

It is possible that the flight path inclination at the end of the
transition is higher than that desired for climb. In this case, the transition has a part where γ is rotated from zero to the climb angle γ_c and a
part where $\gamma = \gamma_c$. The analysis for this case is the same as that for the
landing transition.

6.6.2 Landing transition distance

The landing transition is similar to the take-off transition and is governed by Eqs. (6.37). However, the airplane is assumed to be approaching the runway along the *glide slope* (γ_g = Const) with gear and flaps down. From $h_0 = 50$ ft down, the airplane continues along the glide slope with $n = 1$ until it can switch to the *flare* (n = Const) and land with $h_f = 0, \gamma_f = 0$. The boundary conditions are given by

$$h_0 = 50 \text{ ft}, \quad x_0 = 0, \quad \gamma_0 = \gamma_g$$
$$h_f = 0, \quad \gamma_f = 0. \tag{6.43}$$

Along the glide slope, Eqs. (6.37) can be integrated as

$$x - x_0 = \frac{h - h_0}{\gamma_g}. \tag{6.44}$$

At the point where the switch to the flare is made (subscript s),

$$x_s - x_0 = \frac{h_s - h_0}{\gamma_g}. \tag{6.45}$$

For the flare, the flight path angle is negative so that Eq. (6.39) becomes

$$\gamma = -(1/V)\sqrt{2g(n-1)h} \tag{6.46}$$

and at the switch point gives

$$h_s = \frac{(V\gamma_g)^2}{2g(n-1)}. \tag{6.47}$$

Next the distance equation for the flare can be integrated as

$$x = x_s + \frac{2V}{\sqrt{2g(n-1)}}(\sqrt{h_s} - \sqrt{h}) \tag{6.48}$$

which after application of the final condition becomes

$$x_f = x_s + \frac{2V}{\sqrt{2g(n-1)}}\sqrt{h_s}. \tag{6.49}$$

Finally, combining the results for the glide slope and the flare gives the *landing transition distance*

$$x_f - x_0 = -\frac{50}{\gamma_g} - \frac{V^2\gamma_g}{2g(n-1)} \tag{6.50}$$

6.7   Sample Calculations for the SBJ

6.7.1   Flap aerodynamics: no slats, single-slotted flaps

The K factors for the SBJ (App. A) have been found from Figs. 6.10 through 6.29 and are listed in Table 6.2. Also, the values of ΔC_{L_F}, ΔC_{D_F}, and $C_{L_{max}}$ have been calculated and are given in Table 6.3.

Table 6.2. Values of K Factors for the SBJ

Parameter	K	Parameter	K
$A_W = 5.10$	$K_1 = .88$	$b_S/b_W = 0.0$	$K_{11} = 0.0$
$(t/c)_W = .09$	$K_2 = .95$	$\Lambda_{lew} = 16.5$ deg	$K_{12} = .92$
$A_W = 5.10$	$K_3 = .76$	$R_{N_W} = 8.9510^6$	$K_{13} = 1.0$
$(t/c)_W = .09$	$K_4 = .96$	$M = .2$	$K_{14} = .98$
$c_F/c_W = .178$	$K_5 = .80$	$d_B/b_W = .153$, $c_{r_W}/l_B = .220$	$K_{15} = .01$
$\delta_F = 10$ deg	$K_6 = .43$	$c_F/c_W = .178$	$K_{16} = .50$
$\delta_F = 20$ deg	$K_6 = .73$	$\delta_F/(\delta_F)_r = 10/60 = .17$	$K_{17} = .17$
$\delta_F = 30$ deg	$K_6 = .89$	$\delta_F/(\delta_F)_r = 20/60 = .33$	$K_{17} = .33$
$\delta_F = 40$ deg	$K_6 = .97$	$\delta_F/(\delta_F)_r = 30/60 = .50$	$K_{17} = .50$
$b_F/b_W = .560$	$K_7 = .63$	$\delta_F/(\delta_F)_r = 40/60 = .67$	$K_{17} = .67$
$\Lambda_{qc_W} = 13$ deg	$K_8 = .92$	$b_F/b_W = .560$	$K_{18} = .62$
$c_S/c_W = 0.0$	$K_9 = 0.0$	$\Lambda_F = 5$ deg	$K_{19} = .98$
$\delta_S/(\delta_S)_r = 0.0$	$K_{10} = 0.0$	$d_B/b_W = .153$	$K_{20} = .84$

Table 6.3. Values of ΔC_{L_F}, ΔC_{D_F}, and $C_{L_{max}}$ for the SBJ

δ_F (deg)	ΔC_{L_F}	ΔC_{D_F}	$C_{L_{max}}$
0	0	0	1.157
10	.1732	.00553	1.289
20	.2941	.01106	1.451
30	.3586	.01659	1.516
40	.3908	.02211	1.548

6.7.2   Take-off aerodynamics: $\delta_F = 20$ deg

The aerodynamic quantities needed for take-off are the airplane lift coefficient and the drag coefficient. The formulas are taken from Sec. 6.3 where all the symbols are defined. The lift coefficient is given by

$$C_L = G_L(\bar{h})C_{L_\alpha}(\alpha - \alpha_{0L}) + \Delta C_{L_F}(\delta_F) \tag{6.51}$$

The quantity $\bar{h}$ is the height of the flap trailing edge above the ground. The wing chord plane is 3.34 ft above the ground. The 1.38 ft wide flap is deflected at 20 deg, so $\bar{h} = 2.837$ ft. Next, the angle of attack is taken to be 0.0 deg. The lift coefficient calculation is as follows:

$$C_L = (1.173)(4.08)(1.5/57.3) + .2941 = .4194 \tag{6.52}$$

The drag coefficient is given by

$$\begin{aligned} C_D &= C_{D_0} + \Delta C_{D_{lg}} + \Delta C_{D_F}(\delta_F) \\ &+ G_D(\bar{h})[K/f(\delta_F)][C_L - \Delta C_{L_F}(\delta_F)]^2 \end{aligned} \tag{6.53}$$

In the calculation of the landing gear drag, the take-off gross weight is 13,300 lb. Hence, the drag coefficient calculation is as follows:

$$\begin{aligned} C_D &= .023 + .02746 + .01106+ \\ &(.5096)(.073/.98)(.4194 - .2941)^2 = .06210 \end{aligned} \tag{6.54}$$

6.7.3   Take-off distance at sea level: $\delta_F = 20$ deg

The ground run distance for take-off is given by Eq. (6.31). The values which are not obvious are $W = 13,000$ lb, $\mu = .02$, $C_D - \mu C_L = .05371$, $C_{L_{max}} = 1.451$, $V_{LO} = 1.2(180) = 216$ ft/s and $T = 5,750$ lb. The ground run distance is then calculated to be $x_f = 1,839$ ft. In the transition, the velocity is 216 ft/s and the angle of attack is such that $n = 1.2$. Hence, Eq. (6.42) gives the transition distance to be $x_f = 712.2$ ft. Finally, the take-off distance is the sum of the ground run distance and the transition distance. It is given by $x_f = 2,559$ ft.

6.7.4   Landing aerodynamics: $\delta_F = 40$ deg

The aerodynamic quantities needed for landing are the lift coefficient and the drag coefficient. The formulas are the same as those used for take-off. With the flaps at 40 deg, the height of the trailing edge is $\bar{h} =$ 2.182 ft. The lift coefficient calculation is given by

$$C_L = 1.184(4.08)(1.5/57.3) + .3908 = .5173. \tag{6.55}$$

Similarly, the drag coefficient calculation becomes

$$
\begin{aligned}
C_D &= .023 + .02746 + .02211 \\
&+ .4780(.073/.84)(.5173 - .3908)^2 = .07323.
\end{aligned}
\tag{6.56}
$$

Note that, with $\mu = .35$, $C_D - \mu C_L = -.1078$.

6.7.5   Landing distance at sea level: $\delta_F = 40$ deg

With the glide slope angle of $\gamma_g = 3$ deg, $V=209$ ft/s and $n = 1.2$, Eq. (6.50) gives the transition distance $x_f = 1133$ ft. Then, with a high weight, $W=13,000$ lb, $\mu = .35$, and idle thrust $T=390$ lb, Eq. (6.35) gives the ground run distance of $x_f=2553$ ft. Finally, adding the transition distance and the ground run distance gives a landing distance of $x_f=3686$ ft.

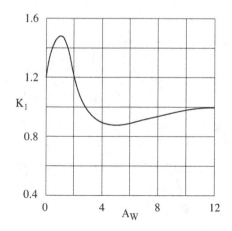

Fig. 6.10: Flap Correction Factor K_1

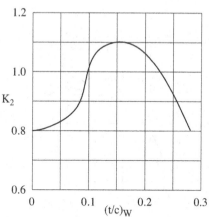

Fig. 6.11: Flap Correction Factor K_2

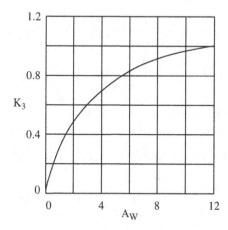

Fig. 6.12: Flap Correction Factor K_3

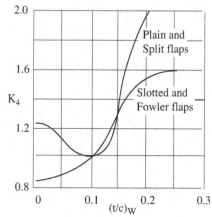

Fig. 6.13: Flap Correction Factor K_4

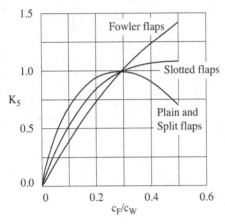

Fig. 6.14: Flap Correction Factor K_5

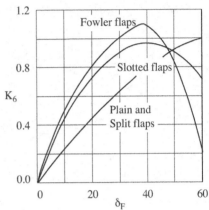

Fig. 6.15: Flap Correction Factor K_6

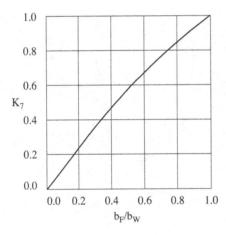

Fig. 6.16: Flap Correction Factor K_7

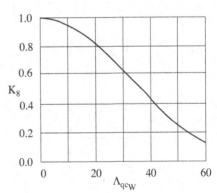

Fig. 6.17: Flap Correction Factor K_8

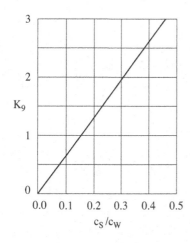

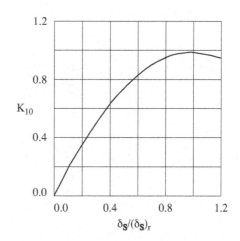

Fig. 6.18: Flap Correction Factor K_9.

Fig 6.19: Flap Correction Factor K_{10}.

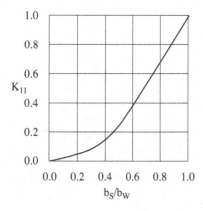

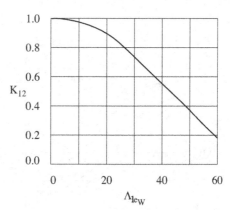

Fig. 6.20: Flap Correction Factor K_{11}.

Fig. 6.21: Flap Correction Factor K_{12}.

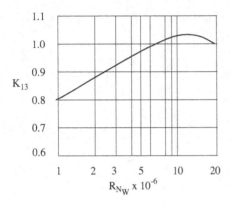

Fig. 6.22: Flap Correction Factor K_{13}

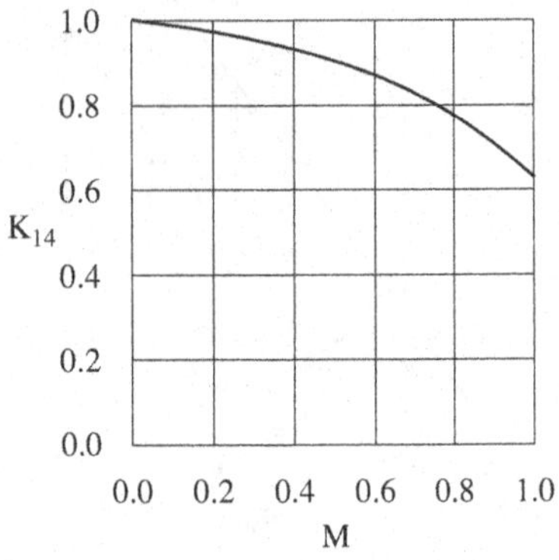

Fig. 6.23: Flap Correction Factor K_{14}.

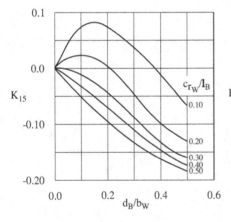

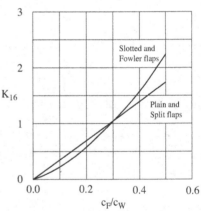

Fig. 6.24: Flap Correction Factor K_{15}.

Fig. 6.25: Flap Correction Factor K_{16}.

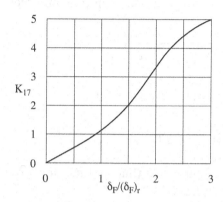

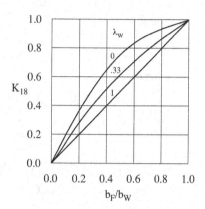

Fig. 6.26: Flap Correction Factor K_{17}

Fig. 6.27: Flap Correction Factor K_{18}

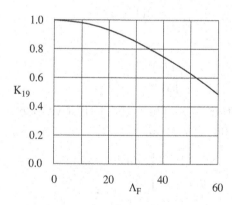

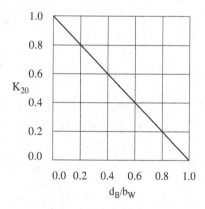

Fig. 6.28: Flap Correction factor K_{19}

Fig. 6.29: Flap Correction Factor K_{20}

Problems

6.1 (a) Derive the equations for the take-off ground run time and fuel.
Show that

$$t_f = \frac{W}{g}\frac{1}{2ab}\ln\frac{a + bV_{LO}}{a - bV_{LO}}$$

where

$$a = \sqrt{T - \mu W}, \quad b = \sqrt{(1/2)\rho S(C_D - \mu C_L)}.$$

Also show that

$$W_0 - W_f = CTt_f$$

(b) For the SBJ (Sec. 6.7.3), calculate the time and fuel for $\delta_F = 20$ deg. Assume that $C = 1.05$ 1/hr.

6.2 What lift coefficient should be designed into an airplane to obtain the minimum take-off ground roll distance?

a. Show that Eq. (6.30) can be rewritten as

$$x_f = -\frac{W}{2g(T - \mu W)}\frac{1}{y}\ln(1 - y)$$

where

$$y = \frac{\rho S(C_D - \mu C_L)V_{LO}^2}{2(T - \mu W)}$$

is the only term that contains C_L. Then, the derivative of x_f with respect to C_L set equal to zero gives

$$\left[\frac{1}{y^2}\ln(1 - y) + \frac{1}{y(1 - y)}\right]\frac{dy}{dC_L} = 0.$$

For typical values of y $(0 < y < .5)$, show that the only solution of this equation is

$$\frac{dy}{dC_L} = 0$$

which with $C_D = C_D(C_L)$ leads to

$$\frac{dC_D}{dC_L} = \mu$$

b. For the drag polar

$$C_D = C_{Dm} + K_m(C_L - C_{Lm})^2,$$

show that the optimal C_L is given by

$$C_L = C_{Lm} + \frac{\mu}{2K_m}.$$

Then, for the take-off ground run of the SBJ (Sec. 6.7.3), show that the optimal value of C_L is C_L=.5575. Since the value of C_L that is designed into the SBJ is C_L=.4194, how might the design be modified to get the optimal C_L (see the first of Eqs. 6.7)?

6.3 Derive the equation for the time during the take-off transition, that is,

$$t_f - t_0 = \sqrt{\frac{2h_f}{g(n-1)}}.$$

6.4 It is desired to determine the change in the velocity during the take-off transition to verify the assumption of velocity constant on the right-hand side of the equations of motion. Prove that $V(h)$ is given by

$$V = -\frac{(a - b\sqrt{h})^2}{b}$$

where

$$a = \frac{g}{W}\frac{T-D}{\sqrt{2g(n-1)}}, \quad b = \frac{g}{V}.$$

Using the boundary conditions

$$h_0 = 0, \quad V_0 = V_{LO}, \quad h_f = 35\text{ft},$$

show that the final velocity becomes

$$V_f = V_{LO} + 2a\sqrt{h_f} - bh_f.$$

Calculate the final velocity for the SBJ in Section 6.7.3. What is the percent change in the velocity over this transition?

6.5 Consider the take-off transition of an airplane at constant load factor n. Show that the airplane is operating at the constant angle of attack

$$\alpha = \alpha_{0L} + \frac{nW/(\bar{q}S) - \Delta C_{L_F}}{C_{L_\alpha}}.$$

Next, show that the angle ϕ that the x_b axis makes with the ground during transition is

$$\phi = \gamma + \alpha = (1/V)\sqrt{2g(n-1)h} + \alpha.$$

At what angle of attack is the SBJ flying (Sec. 6.7.3)? What is the inclination of the x_b axis with respect to the ground during transition.

6.6 Consider the take-off transition for the case where the flight path angle is rotated from $\gamma = 0$ to the climb value γ_c which is held constant over the remainder of the transition. Derive the equations for the distance and the time, that is,

$$x_f = \frac{V^2\gamma_c}{2g(n-1)} + \frac{h_f}{\gamma_c}, \quad t_f = \frac{V\gamma_c}{2g(n-1)} + \frac{h_f}{V\gamma_c}.$$

6.7 Analyze the effect of flaps on the take-off ground run distance of the SBJ. Compute the take-off distance for $\delta_F = 0, 10, 20, 30$, and 40 deg recalling that the lift-off speed changes with $C_{L_{max}}$ which changes with δ_F. If the take-off distance always decreases as δ_F increases, why do you think that $\delta_F = 20$ deg should be used for take-off?

6.8 It is desired to find the constant thrust and the parabolic thrust approximations a thrust table. For the two SBJ turbojets, thrust versus velocity can be represented by the following table:

V (ft/s)	Thrust (lb)
0	6,000
122	5,740
246	5,500

a. For the constant thrust approximation, there are more points than unknowns. Hence, the fit is accomplished by minimizing the sum of the squares of the errors, that is,

$$J = (T - T_1)^2 + (T - T_2)^2 + (T - T_3)^2.$$

with respect to T. Here, T is the desired constant thrust, and T_1, T_2, T_3 are the values of the thrust at V_1, V_2, V_3. Show that

$$T = \frac{T_1 + T_2 + T_3}{3}.$$

Show that the constant thrust is T=5,750 lb.

b. Find the parabola $T = PV^2 + QV + R$ that fits the data. Show that

$$P = .0007953, \quad Q = -2.228, \quad R = 6,000.$$

6.9 Consider the take-off ground run for the case where the thrust is modeled as
$$T = PV^2 + QV + R.$$

a. Show that the ground run distance integral is given by

$$x_f = -\frac{W}{g} \int_0^{V_{LO}} \frac{V\,dV}{aV^2 + bV + c}$$

where

$$a = -P + (1/2)\rho S(C_D - \mu C_L)$$
$$b = -Q$$
$$c = -R + \mu W$$

b. Assuming that $b^2 - 4ac > 0$, show that the ground run distance is given by

$$x_f = -(W/g)[A(V_{LO}) - A(0)]$$

where

$$A(V) = \frac{1}{2a} \ln \left| aV^2 + bV + c \right| - \frac{b}{2aq} \ln \left| \frac{2aV + b - q}{2aV + b + q} \right|$$

where

$$q = \sqrt{b^2 - 4ac}.$$

c. The thrust for the two SBJ turbojets can be modeled as

$$T = .0007953V^2 - 2.228V + 6,000.$$

For the take-off ground run with $\delta_F = 20$ deg (Sec. 6.7.3),
compute a, b, c, q and compute the ground distance, that is,

$$a = .01400, \quad b = 2.228, \quad c = -5,740, \quad q = 18.1, \quad x_f = 2,188 ft.$$

How does this value of the ground run distance compare with
that obtained with constant thrust (Sec. 6.7.3)?

6.10 On take-off, a multi-engine airplane can lose an engine during the
ground run. Depending on the value of the decision speed, the
airplane leaves the ground and goes around to land or the airplane
is braked to a stop. The decision speed is the speed from which the
airplane can be stopped with brakes only and zero thrust within
the length of the runway. Derive the formula for calculating the decision speed, and calculate it for the SBJ in Section 6.7.3 assuming
the field length is 6,000 ft. What does this result mean?

6.11 Derive the equation for the time during the landing transition, that
is,

$$t_f - t_0 = -\frac{h_0}{V\gamma_g} - \frac{V\gamma_g}{2g(n-1)}$$

6.12 Derive the equation for the time during the landing ground run.

6.13 Consider the effect of thrust reversal on landing distance. Assume
that the thrust is reversed $(T = T_r < 0)$ at a particular speed
V_r during the ground run. Derive the formula for the ground run
distance.

6.14 For a fighter aircraft that is equipped with thrust vector control,
the thrust magnitude T and the thrust direction relative to the
airplane ε_0 can both be controlled. For take-off, it is desired to
find the angle ε_0 which minimizes the take-off ground run distance.
Since the angle of attack is constant, minimizing with respect to
ε_0 is the same as minimizing with respect to ε. In this analysis,
the assumption $T \sin \varepsilon \ll W$ is not made.

a. Assume that the thrust angle of attack ε is constant during
the ground run and that V_{LO} is specified. Show that the
ground run distance is given by

$$x_f = -\frac{W}{g\rho S(C_D - \mu C_L)} \ln \left[1 - \frac{\rho S(C_D - \mu C_L)V_{LO}^2}{2(T\cos\varepsilon - \mu W + \mu T\sin\varepsilon)} \right].$$

b. Show that the thrust inclination that optimizes the ground run distance is given by

$$\tan \varepsilon = \mu$$

which for $\mu = 0.02$ is very small.

c. Show that the minimal ground run distance is given by

$$x_f = -\frac{W}{g\rho S(C_D - \mu C_L)} \ln \left[1 - \frac{\rho S(C_D - \mu C_L)V_{LO}^2}{2(T\sqrt{1+\mu^2} - \mu W)} \right]$$

Note that since $\mu^2 << 1$ the ground run distance reduces to that for $\varepsilon = 0$, that is, Eq. (6.31).

6.15 Relative to Prob. 6.14 assume that ε is constant but that the lift-off speed is not prescribed. It is assumed that D and L do not vary with ε.

a. Starting from Eqs. (6.17), show that the stall speed $(R = 0)$ is given by

$$V_{stall} = \sqrt{\frac{2W(1 - \tau \sin \varepsilon)}{\rho S C_{Lmax}}}$$

where $\tau = T/W$ is the thrust to weight ratio.

b. Next, show that the take-off ground run distance satisfies the equation

$$x_f = -\frac{W}{g\rho S(C_D - \mu C_L)} \ln \left[1 - \frac{(C_D - \mu C_L)(k^2/C_{Lmax})(1 - \tau \sin \varepsilon)V_{TD}^2}{\tau \cos \varepsilon - \mu(1 - \tau \sin \varepsilon)} \right].$$

c. Show that the optimal thrust angle is given by

$$\sin \varepsilon = \tau.$$

Discuss the meaning of this result as τ varies between 0 and 1.

d. Show that the optimal ground run distance is given by

$$x_f = -\frac{W}{g\rho S(C_D - \mu C_L)} \ln \left[1 - \frac{(C_D - \mu C_L)(k^2/C_{Lmax})\sqrt{1 - \tau^2}}{\tau - \mu\sqrt{1 - \tau^2}} \right]$$

where τ is the thrust to weight ratio T/W. Note that if $\tau^2 << 1$, the ground run distance reduces to that of Eq. (6.31). Hence, there is no advantage to inclining the thrust for τ's less than around $\tau = 0.3$ or so. What is the value of τ for the SBJ?

6.16 Consider the problem of minimizing the landing ground run distance with brakes and reverse thrust applied all the way from $V_0 = V_{TD}$ to $V_f = 0$ where V_{TD} is given. Let ε denote the inclination of the forward-pointing engine centerline. With $T < 0$, the thrust vector is actually pointing rearward. The assumption that $T \sin \varepsilon << W$ is not made here.

a. Show that the landing ground run distance is given by

$$x_f = -\frac{W}{g\rho S(C_D - \mu C_L)} \ln\left[1 - \frac{\rho S(C_D - \mu C_L)V_{TD}^2}{2(T\cos\varepsilon - \mu W + \mu T \sin\varepsilon)}\right].$$

where $C_D - \mu C_L < 0$.

b. Show that

$$\tan\varepsilon = \mu$$

minimizes the ground run distance. This result says that the reverse thrust vector has a component into the ground. Does it make sense?

6.17 Reconsider Prob. 6.16 for the case where V_{TD} is not given, but satisfies the relation

$$V_{TD} = kV_{stall} = k\sqrt{\frac{2W(1 - \tau\sin\varepsilon)}{\rho S C_{L_{max}}}}.$$

where $\tau = T/W$ is the thrust to weight ratio. Note that, since $\tau < 0$, a positive ε increases the touchdown speed.

a. Show that the ground run distance is given by

$$x_f = -\frac{W}{g\rho S(C_D - \mu C_L)} \ln\left[1 - \frac{(C_D - \mu C_L)(k^2/C_{L_{max}})(1 - \tau\sin\varepsilon)V_{TD}^2}{\tau\cos\varepsilon - \mu(1 - \tau\sin\varepsilon)}\right].$$

b. Show that the optimal thrust angle is given by

$$\sin\varepsilon = \tau.$$

Since $\tau < 0$ ($T < 0$), this means that the reverse thrust vector has an upward component (away from the ground). Does this result make sense?

Chapter 7

P_S and Turns

In the military aircraft mission profile (Chap. 1), there are three additional segments: acceleration at constant altitude, specific excess power P_S, and turns. Each of these segments involves a nonzero dynamic term. Level flight acceleration (dash) is not considered here because the aircraft must accelerate through the transonic region, and no analytical solution can be obtained. Acceleration from one subsonic speed to another is given as a homework problem. Specific excess power is a measure of the ability of the airplane to change energy. It is considered first because the equations of motion already exist. For turns, new equations must be derived. Turns are used to estimate the amount of fuel needed for combat maneuvering in the area of the target, as well as to change the heading of an airplane.

7.1 Accelerated Climb

A high-performance airplane increases its speed during the climb. Hence, to analyze the climb performance of such an airplane, it is necessary to include the $\dot{V}$ term in the equations of motion. To discuss an accelerated climb, the following problem is studied: For a given power setting, find the climb schedule that minimizes the time to climb from a given initial altitude to a given final altitude.

In order to analyze this problem, the standard assumptions of small thrust angle of attack, negligible thrust component normal to the flight path, and small normal acceleration are made. In addition, the weight is assumed to be constant since the weight only changes 5-10% during the climb. With these approximations, the equations of motion for flight in a vertical plane reduce to

$$\dot{x} = V \cos \gamma$$
$$\dot{h} = V \sin \gamma$$
$$\dot{V} = (g/W)[T(h, V, P) - D(h, V, L) - W \sin \gamma] \qquad (7.1)$$
$$0 = L - W \cos \gamma$$
$$\dot{W} = -C(h, V, P)T(h, V, P) .$$

These equations have two mathematical degrees of freedom. Since the power setting is held constant, the remaining degree of freedom is associated with the velocity.

Because the lift is not equal to the weight, these equations cannot be solved analytically. The accelerated climb problem can be solved analytically if the drag can be written as $D(h, V, W)$. This can be accomplished in two ways. The first is to assume small flight path inclination ($\cos \gamma = 1$, $\sin \gamma = \gamma$) so that $L = W$. However, high performance airplanes can climb at high values of γ. The second approach is to assume that that part of the drag which comes from $L \neq W$ is negligible with respect to the remainder. To see this, consider the parabolic drag expression

$$D = (1/2)C_{D_0}\rho SV^2 + 2KL^2/(\rho SV^2) \qquad (7.2)$$

From Eqs. (7.1) it is seen that

$$L^2 = W^2 \cos^2 \gamma = W^2 - W^2 \sin^2 \gamma \qquad (7.3)$$

so that the drag can be rewritten as

$$D = (1/2)C_{D_0}\rho SV^2 + 2KW^2/(\rho SV^2)$$
$$-2KW^2 \sin^2 \gamma/(\rho SV^2). \qquad (7.4)$$

The next step is to show that the third term is negligible with respect to the sum of the first two, that is,

$$\frac{2KW^2 \sin^2 \gamma/(\rho SV^2)}{(1/2)C_{D_0}\rho SV^2 + 2KW^2/(\rho SV^2)} \ll 1. \qquad (7.5)$$

This expression can be rewritten as

$$\frac{\sin^2 \gamma}{1 + (V/V^*)^4} \ll 1 \tag{7.6}$$

where V^* is the speed for maximum lift-to-drag ratio. For comparison purposes, consider the small flight path inclination assumption,

$$\cos \gamma = \sqrt{1 - \sin^2 \gamma} \cong 1 - (1/2) \sin^2 \gamma. \tag{7.7}$$

Hence, for the cosine to be approximated by unity, it is necessary that

$$\frac{\sin^2 \gamma}{2} \ll 1. \tag{7.8}$$

Since minimum-time climb speeds in quasi-steady flight are such that are such that $V > V^*$, approximation (7.6) is less restrictive than approximation (7.8). In either case, however, it is possible to write $D = D(h, V, W)$. For a parabolic drag polar, this means that the drag is approximated by

$$D = (1/2)C_{D_0}\rho S V^2 + 2KW^2/(\rho S V^2). \tag{7.9}$$

which is the drag for $L = W$.

The altitude and velocity equations are now given by

$$\begin{aligned} \dot{h} &= V \sin \gamma \\ \dot{V} &= (g/W)[T(h, V, P) - D(h, V, W) - W \sin \gamma] \end{aligned} \tag{7.10}$$

If $\sin \gamma$ is eliminated between these two equations, the resulting equation can be solved for dt and integrated to give

$$t_f - t_0 = \int_{h_0}^{h_f} \frac{1 + (V/g)V'}{P_S(h, V, W, P)} dh \tag{7.11}$$

The symbol V' denotes the derivative dV/dh, and the initial and final altitudes are given. Finally, P_S denotes the *specific excess power* which is defined as

$$P_S = \left[\frac{T(h, V, P) - D(h, V, W)}{W} \right] V \tag{7.12}$$

Note that P_S is the *thrust power TV* minus the *drag power DV* divided by the weight.

With regard to the power setting, the minimum time climb occurs with maximum P. With respect to $V(h)$, Eq. (7.11) has the following functional form:

$$t_f - t_0 = \int_{h_0}^{h_f} f[h, V(h), V'(h)]dh \qquad (7.13)$$

The determination of the function $V(h)$ which minimizes this integral is a problem of the calculus of variations/optimal control theory. It is beyond the scope of this text.

7.2 Energy Climb

While the original problem cannot be solved here, it is possible to transform it into a form which can be solved. In this connection, the *specific energy*, that is, the energy per unit weight,

$$E_S = \frac{mgh + \frac{1}{2}mV^2}{mg} = h + \frac{V^2}{2g} \qquad (7.14)$$

is introduced. From this expression, it is seen that

$$dE_S = [1 + (V/g)V']dh. \qquad (7.15)$$

Hence, if E_S is made the variable of integration, the time to climb (7.11) can be rewritten as

$$t_f - t_0 = \int_{E_{S_0}}^{E_{S_f}} \frac{dE_S}{P_S(h, V, W, P)} \qquad (7.16)$$

where

$$h = E_S - V^2/2g. \qquad (7.17)$$

Note that this integral has the form from which it is easy to find the velocity profile $V(E_S)$ which minimizes the time to climb between the initial energy E_{S_0} and the final energy E_{S_f}. This $V(E_S)$ is obtained by finding the velocity which maximizes P_S at each value of E_S. Mathematically, the $V(E_S)$ which minimizes $t_f - t_0$ is obtained from the condition

$$\left. \frac{\partial P_S}{\partial V} \right|_{E_S = \text{Const}} = 0. \qquad (7.18)$$

From Eq. (7.14) and (7.1), it can be shown that

$$\dot{E}_S = P_S, \tag{7.19}$$

meaning that maximum P_S is the same as maximum $\dot{E}_S$ (maximum energy rate). Hence, the optimal climb schedule is that of maximum energy rate. Recall that the optimal quasi-steady climb is flown at maximum altitude rate (maximum rate of climb).

To solve Eq. (7.18), P_S can be plotted versus V for given values of E_S as in Fig. 7.1. The locus of points at which P_S is a maximum yields $V(E_S)$ for the minimum-time climb.

It can be shown that the optimal velocity profile for climbing from one altitude to another is the same as the optimal velocity profile for climbing from one specific energy to another.

7.3   The P_S Plot

Another way to look at this problem is to plot contours of $P_S = \text{Const}$ in the V, h plane as shown in Fig. 7.2. This plot is known as the P_S plot. Note that the $P_S = 0$ curve is the unrestricted, level flight envelope. Finally, the optimal nonsteady climb (energy climb) is given by the locus of points obtained by maximizing P_S along an $E_S = \text{Const}$ line. Recall that the optimal quasi-steady climb $V(h)$ is the locus of points obtained by maximizing P_S along an $h = \text{Const}$ line.

7.4   Energy Maneuverability

The equations of motion for three-dimensional flight are listed in Chap. 2. If it is assumed that ε is small, the $\dot{h}$ and $\dot{V}$ equations can be combined to give $\dot{E}_S = P_S(h, V, W, P, L)$ since $L \neq W$. In terms of the load factor $n = L/W$, this expression can be rewritten as

$$\dot{E}_S = P_S(h, V, W, P, n) \tag{7.20}$$

where

$$P_S = \left[\frac{T(h, V, P) - D(h, V, W, n)}{W} \right] V. \tag{7.21}$$

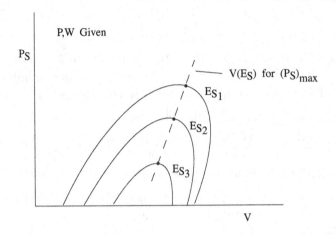

Figure 7.1: Specific Excess Power Versus Velocity

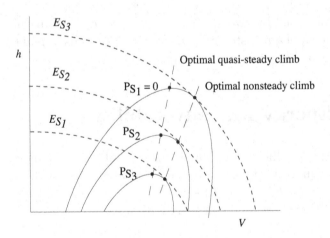

Figure 7.2: The P_S Plot

A typical plot of P_S versus Mach Number for a high-performance maneuvering aircraft is shown in Fig. 7.3.

The ability of an aircraft to change its energy in three-dimensional maneuvering flight is determined by its P_S. Given two fighters engaged in combat, the fighter with the higher P_S will be able to out maneuver the other.

In designing a new fighter, it is necessary to have a P_S higher than that of the expected threat. In analyzing an existing fighter, it is useful to know the region of the flight envelope where it has a higher P_S than that of a potential threat. Then, air combat should only be attempted at these values of h and V.

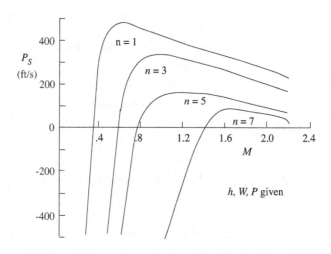

Figure 7.3: P_S Plot for a Maneuvering Aircraft

7.5 Nonsteady, Constant Altitude Turns

One purpose of the turn is to change the heading of the airplane. Turns occur in many parts of a typical flight path. On take-off, the pilot may perform a climbing turn in order to line up with the heading at which the climb to altitude is to be made. In flying along controlled airways, it is often necessary to change from one heading to another in order to change airways. These turns are made at constant altitude or in the horizontal plane. Only turns in a horizontal plane are considered here. Another purpose of the turn is to estimate the fuel required for air combat. This is done by computing the fuel needed to perform a given number of subsonic and supersonic turns at constant altitude (horizontal turns).

In this study, only *coordinated turns* are considered, that is, turns with zero sideslip angle where the velocity vector is always in the plane of symmetry of the airplane. As a consequence, thrust, drag, and lift are also in the aircraft plane of symmetry.

Fig. 7.4 shows the coordinate systems to be used in the derivation of the equations of motion. The ground axes system $Exyz$ is shown, but it is not in the plane of the turn. The x axis is in the original direc-

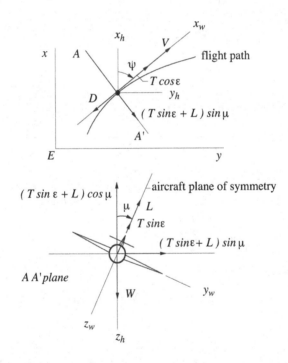

Figure 7.4: Nomenclature for Turning Flight

tion of motion. The subscript h refers to the local horizon system, and the subscript w denotes the wind axes system. Also, ψ is the *heading angle*, and μ is the *bank angle*. Motion of the airplane is restricted to a horizontal plane.

The kinematic equations follow from the definition of velocity (see Chap. 2), that is,

$$\mathbf{V} = d\mathbf{EO}/dt \tag{7.22}$$

where $\mathbf{EO}$ is the position vector of the airplane relative to the ground. It is given by

$$\mathbf{EO} = x\mathbf{i} + y\mathbf{j} - h\mathbf{k} = x\mathbf{i}_h + y\mathbf{j}_h - h\mathbf{k}_h. \tag{7.23}$$

Since $\mathbf{i}_h, \mathbf{j}_h, \mathbf{k}_h$, and h are constant,

$$d\mathbf{EO}/dt = \dot{x}\mathbf{i}_h + \dot{y}\mathbf{j}_h. \tag{7.24}$$

With the velocity vector expressed as

$$\mathbf{V} = u\mathbf{i}_h + v\mathbf{j}_h \tag{7.25}$$

equating like components in Eqs. (7.24) and (7.25) leads to the kinematic equations of motion

$$\dot{x} = u, \quad \dot{y} = v. \tag{7.26}$$

The application of $\mathbf{F} = m\mathbf{a}$ yields the dynamic equations. In terms of the individual forces acting on the vehicle, Newton's law is written as

$$\mathbf{T} + \mathbf{D} + \mathbf{L} + \mathbf{W} = m\mathbf{a} \tag{7.27}$$

where $\mathbf{a}$ is the acceleration of the airplane relative to the ground axes system which is an approximate inertial frame. From Fig. 7.4 it is seen that the resultant force acting on the airplane is given by

$$\begin{aligned}
\mathbf{F} = &\ [(T\cos\varepsilon - D)\cos\psi - (T\sin\varepsilon + L)\sin\mu\sin\psi]\mathbf{i}_h \\
+ &\ [(T\cos\varepsilon - D)\sin\psi + (T\sin\varepsilon + L)\sin\mu\cos\psi]\mathbf{j}_h \tag{7.28} \\
+ &\ [W - (T\sin\varepsilon + L)\cos\mu]\mathbf{k}_h
\end{aligned}$$

From the definition of acceleration

$$\mathbf{a} = d\mathbf{V}/dt \tag{7.29}$$

and the expression (7.25) for $\mathbf{V}$, the acceleration becomes

$$\mathbf{a} = \dot{u}\mathbf{i}_h + \dot{v}\mathbf{j}_h. \tag{7.30}$$

Combining Eqs. (7.28) and (7.30) leads to the dynamic equations

$$
\begin{aligned}
(W/g)\dot{u} &= (T\cos\varepsilon - D)\cos\psi - (T\sin\varepsilon + L)\sin\mu\sin\psi \\
(W/g)\dot{v} &= (T\cos\varepsilon - D)\sin\psi + (T\sin\varepsilon + L)\sin\mu\cos\psi \qquad (7.31) \\
0 &= W - (T\sin\varepsilon + L)\cos\mu
\end{aligned}
$$

where

$$
V = \sqrt{u^2 + v^2}, \quad \tan\psi = v/u. \qquad (7.32)
$$

The equations of motion are the kinematic equations (7.26), the dynamic equations (7.31), and the weight equation $\dot{W} = -CT$. In the dynamic equations, the thrust angle of attack is related to the airplane angle of attack as

$$
\varepsilon = \varepsilon_0 + \alpha. \qquad (7.33)
$$

The equations of motion can be expressed in terms of the velocity V and the heading angle ψ by writing

$$
u = V\cos\psi, \quad v = V\sin\psi. \qquad (7.34)
$$

The corresponding equations of motion for nonsteady turning flight in a horizontal plane are given by

$$
\begin{aligned}
\dot{x} &= V\cos\psi \\
\dot{y} &= V\sin\psi \\
\dot{V} &= (g/W)[T\cos(\varepsilon_0 + \alpha) - D] \\
\dot{\psi} &= (g/WV)[T\sin(\varepsilon_0 + \alpha) + L]\sin\mu \qquad (7.35) \\
0 &= [T\sin(\varepsilon_0 + \alpha) + L]\cos\mu - W \\
\dot{W} &= -CT
\end{aligned}
$$

In these equations, the angle of attack, the drag, the thrust, and the specific fuel consumption satisfy functional relations of the form

$$
\begin{aligned}
\alpha &= \alpha(h, V, L), \quad D = D(h, V, L) \\
T &= T(h, V, P), \quad C = C(h, V, P). \qquad (7.36)
\end{aligned}
$$

7.6 Quasi-Steady Turns: Arbitrary Airplane

To study quasi-steady turns, the following approximations are made: (a) $\dot{V}$ negligible, (b) thrust angle of attack small, (c) negligible component of the thrust normal to the flight path, and (d) weight constant on the right hand sides of the equations of motion. With these approximations, the equations of motion become

$$\dot{x} = V \cos \psi \tag{7.37}$$

$$\dot{y} = V \sin \psi \tag{7.38}$$

$$0 = T(h, V, P) - D(h, V, L) \tag{7.39}$$

$$\dot{\psi} = gL \sin \mu / WV \tag{7.40}$$

$$0 = L \cos \mu - W \tag{7.41}$$

$$\dot{W} = -C(h, V, P)T(h, V, P) \ . \tag{7.42}$$

These six equations involve eight variables (four states x, y, ψ, W and four controls V, P, L, μ) and, hence, two mathematical degrees of freedom. If it is assumed that the turn is flown at constant power setting, there is only one mathematical degree of , and it is associated with the velocity.

Since the altitude, weight, and power setting are given, Eq. (7.39) can be solved for the *load factor* ($n = L/W$) in terms of the velocity as

$$n = n(V). \tag{7.43}$$

Then, Eq. (7.41) can be solved for the *bank angle* as

$$\cos \mu = \frac{1}{n} \tag{7.44}$$

so that

$$\sin \mu = \sqrt{1 - \left(\frac{1}{n}\right)^2} = \frac{\sqrt{n^2 - 1}}{n} \ . \tag{7.45}$$

Hence, $\mu = \mu(V)$.

The objective of this study is to determine the distance, time and fuel consumed during a turn from one heading angle to another.

Hence, ψ is made the variable of integration. Before continuing, it is noted that the *turn rate* $\dot{\psi}$ is given by Eq. (7.40) can be rewritten as

$$\dot{\psi} = \frac{g\sqrt{n^2 - 1}}{V} \qquad (7.46)$$

so that $\dot{\psi} = \dot{\psi}(V)$.

To determine the distance along the turn, the arc length along an infinitesimal section of the turn is given by $ds^2 = dx^2 + dy^2$. Taking the square root and integrating with respect to time leads to

$$s_f - s_0 = \int_{t_0}^{t_f} \sqrt{\dot{x}^2 + \dot{y}^2}\, dt \ . \qquad (7.47)$$

Then, from Eqs. (7.37) and (7.38) and switching to ψ as the variable of integration, the distance along the turn becomes

$$s_f - s_0 = \int_{\psi_0}^{\psi_f} \frac{V}{\dot{\psi}(V)}\, d\psi \ . \qquad (7.48)$$

Note that the integrand is the instantaneous *turn radius*

$$r = \frac{V}{\dot{\psi}(V)} = \frac{V^2}{g\sqrt{n^2 - 1}}. \qquad (7.49)$$

Next, the time to make the turn is given by

$$t_f - t_0 = \int_{\psi_0}^{\psi_f} \frac{d\psi}{\dot{\psi}(V)}, \qquad (7.50)$$

and the fuel consumed is

$$W_o - W_f = \int_{\psi_0}^{\psi_f} \frac{C(V)T(V)}{\dot{\psi}(V)}\, d\psi. \qquad (7.51)$$

Because the integrands of the distance, time and fuel contain only the velocity and not the variable of integration, all of the optimal turns are made at constant velocity. The minimum distance turn is flown at the speed for minimum radius; the minimum time turn is flown at the speed for maximum turn rate; and the minimum fuel turn is flown at the speed for minimum $CT/\dot{\psi}$. Note that if C and T are independent of V, the minimum fuel turn and the minimum time turn are flown at the same speed, that for maximum turn rate.

For a constant velocity turn, the distance, time, and fuel integrals can be evaluated to obtain

$$s_f - s_0 \;=\; r(V)(\psi_f - \psi_0)$$

$$t_f - t_0 \;=\; \tfrac{1}{\dot\psi(V)}(\psi_f - \psi_0) \tag{7.52}$$

$$W_o - W_f \;=\; \tfrac{C(V)T(V)}{\dot\psi(V)}(\psi_f - \psi_0)\,.$$

Finally, since the radius of the turn is constant, the trajectory is a circle centered on the positive y axis. Recall that these results are for an arbitrary conventional airplane (airframe and engines) flying a constant altitude turn at constant velocity.

The turn performance has been computed for the subsonic business jet (SBJ) of App. A. Results are presented for $W = 11{,}000$ lb, $P = 0.98$ (maximum continuous thrust), and several values of the altitude ($h = 0$, 20,000, and 40,000 ft). First, the load factor, the bank angle, the turn rate, and the turn radius are shown in Figs. 7.5 through 7.8. The fuel is not shown because it is expected to look like the turn rate. Next, the optimal speeds for the minimum time and distance turns are shown in the flight envelope in Fig. 7.9. Finally, the maximum turn rate and minimum turn radius are plotted in Figs. 7.10 and 7.11

7.7 Flight Limitations

In addition to the dynamic pressure, Mach number, and lift coefficient limits shown in Fig. 7.9, there is also a structural design limit on the load factor. While commercial jets might have a load factor limit of 2.5, modern jet fighters are designed for a maximum load factor of 9.0. For the SBJ, the maximum dynamic pressure is 300 lb/ft^2; the maximum Mach number is .81; the maximum lift coefficient is 1.24; and the maximum load factor is 4.5. Note that at sea level the SBJ is able to fly at a load factor which exceeds the limit (Fig. 7.5).

From Eqs. (7.46) and (7.49), it is seen that the turn rate and the turn radius are given by

$$\dot\psi = \frac{g\sqrt{n^2-1}}{V}, \quad r = \frac{V^2}{g\sqrt{n^2-1}} \tag{7.53}$$

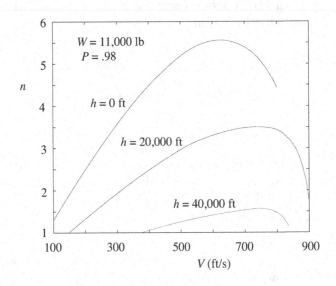

Figure 7.5: Load Factor (SBJ)

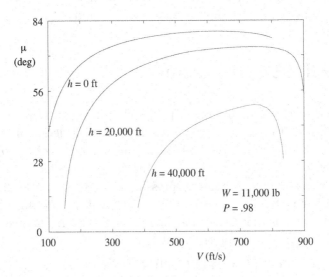

Figure 7.6: Bank Angle (SBJ)

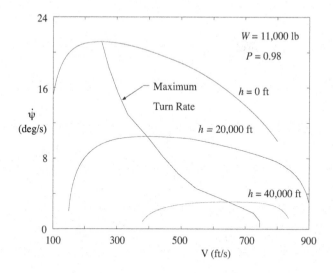

Figure 7.7: Turn Rate (SBJ)

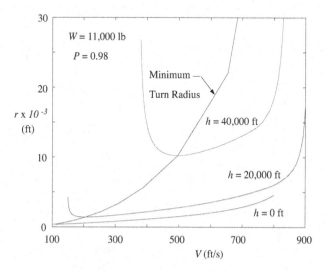

Figure 7.8: Turn Radius (SBJ)

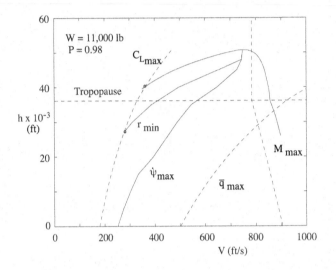

Figure 7.9: Optimal Flight Speeds (SBJ)

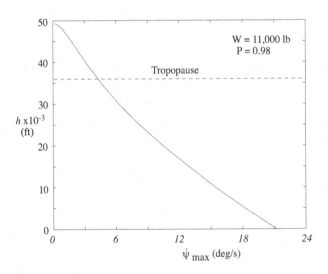

Figure 7.10: Maximum Turn Rate versus Altitude

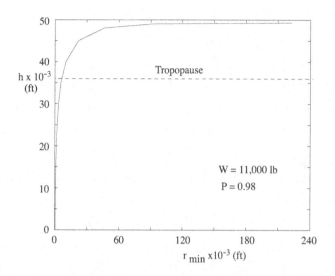

Figure 7.11: Minimum Radius versus Altitude (SBJ)

where

$$n = \frac{L}{W} = \frac{C_L \rho S V^2}{2W} = \frac{C_L}{C_{L_{max}}} \left(\frac{V}{V_{stall}}\right)^2. \tag{7.54}$$

Hence, for the turn rate the limit lift coefficient ($C_{L_{max}}$) curve and the limit load factor curve are defined by the relations

$$\begin{aligned} \dot{\psi} &= \frac{g\sqrt{(V/V_{stall})^4 - 1}}{V} \\ \dot{\psi} &= \frac{g\sqrt{n_{lim}^2 - 1}}{V}. \end{aligned} \tag{7.55}$$

For the turn radius, the corresponding curves are given by

$$\begin{aligned} r &= \frac{V^2}{g\sqrt{(V/V_{stall})^4 - 1}} \\ r &= \frac{V^2}{g\sqrt{n_{lim}^2 - 1}}. \end{aligned} \tag{7.56}$$

The turn rate limits at $h = 20{,}000$ ft have been added to Fig. 7.7 and are shown in Fig. 7.12. The speed at which the limit C_L curve and the limit load factor curve intersect is called the *corner speed* V_c and is given by

$$V_c = \sqrt{n_{lim}}\, V_{stall} = \sqrt{n_{lim}}\, \sqrt{\frac{2W}{\rho S C_{L_{max}}}}. \tag{7.57}$$

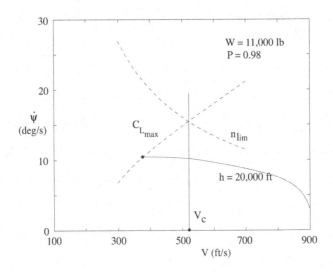

Figure 7.12: Corner Speed (SBJ)

Note that the corner speed increases with altitude.

If the turn rate capability exceeds the limit as it does at sea level, the quickest turn is achieved by reducing the power setting and flying at the corner speed. It is not unusual for fighters to exceed the turn limits over a wide range of altitudes. Here, these airplanes achieve the quickest and tightest turn by flying at the corner speed.

7.8   Quasi-steady Turns: Ideal Subsonic Airplane

Since optimal turn performance of the SBJ occurs at speeds where Mach number effects are negligible, the drag polar can be approximated by a parabolic drag polar with constant coefficients, that is,

$$C_D = C_{D_0} + K C_L^2 \qquad (7.58)$$

where C_{D_0} and K are constants. The approximate thrust and specific fuel consumption of the subsonic business jet can be modeled as (see

Chap. 3)

$$T = T(h, P), \quad C = C(h). \tag{7.59}$$

For the parabolic drag polar, the expression for the drag is given by

$$D = (1/2)C_{D_0}\rho SV^2 + 2KL^2/(\rho SV^2) \tag{7.60}$$

or because of the definition of the load factor, by

$$D = (1/2)C_{D_0}\rho SV^2 + 2KW^2n^2/(\rho SV^2). \tag{7.61}$$

Next, the nondimensional velocity

$$u = V/V^* \tag{7.62}$$

is introduced where the velocity for maximum lift-to-drag ratio is given by

$$V^* = (2W/\rho SC_L^*)^{1/2}, \quad C_L^* = (C_{D_0}/K)^{1/2}. \tag{7.63}$$

Then, the drag can be rewritten as

$$D = (W/2E^*)(u^2 + n^2/u^2) \tag{7.64}$$

For level flight ($n = 1$), the quantity W/E^* is the minimum drag for the given altitude and weight. Hence, it is the minimum thrust required to fly at constant altitude at the given altitude and weight, that is,

$$T_{min} = W/E^*. \tag{7.65}$$

Since the altitude and power setting are given, the thrust is fixed, and a nondimensional thrust is introduced as

$$\tau = T/T_{min}. \tag{7.66}$$

The *load factor* is obtained from the equation $T - D = 0$ which can be written in terms of the nondimensional variables as

$$\tau - (1/2)(u^2 + n^2/u^2) = 0. \tag{7.67}$$

The load factor is then given by

$$n = u\sqrt{2\tau - u^2}, \tag{7.68}$$

and the *bank angle* is given by

$$\mu = \arccos \frac{1}{u\sqrt{2\tau - u^2}} . \qquad (7.69)$$

From Eq. (7.53), it is seen that the *turn rate* in terms of the nondimensional speed is given by

$$\frac{\dot\psi V^*}{g} = \sqrt{2\tau - u^2 - 1/u^2}. \qquad (7.70)$$

Then, the turn rate has a maximum of

$$\frac{\dot\psi V^*}{g} = \sqrt{2(\tau - 1)} \qquad (7.71)$$

at the velocity

$$u = 1. \qquad (7.72)$$

The *turn radius* (7.53) can be written in terms of the nondimensional speed as

$$\frac{rg}{V^{*2}} = \frac{u^2}{\sqrt{2\tau u^2 - u^4 - 1}} \qquad (7.73)$$

and has the minimum value

$$\frac{rg}{V^{*2}} = \frac{V^{*2}}{g} \frac{1}{\sqrt{\tau^2 - 1}} \qquad (7.74)$$

when

$$u = \frac{1}{\sqrt{\tau}}. \qquad (7.75)$$

The C_L and n limit curves follow from Eqs, (7.53) and (7.54). The limit curves for the turn rate are given by

$$\begin{aligned}
\frac{\dot\psi V^*}{g} &= \frac{\sqrt{(C_{Lmax}/C_L^*)^2 u^4 - 1}}{u} \\
\frac{\dot\psi V^*}{g} &= \frac{g\sqrt{n_{lim}^2 - 1}}{u},
\end{aligned} \qquad (7.76)$$

and the limit curves for the turn radius become

$$\begin{aligned}
\frac{rg}{V^{*2}} &= \frac{u^2}{\sqrt{(C_{Lmax}/C_L^*)^2 u^4 - 1}} \\
\frac{rg}{V^{*2}} &= \frac{u^2}{g\sqrt{n_{lim}^2 - 1}}.
\end{aligned} \qquad (7.77)$$

Finally, the *corner speed* is given by

$$u_c = \sqrt{n_{lim}}\sqrt{C_L^*/C_{L_{max}}} \; . \tag{7.78}$$

Problems

7.1 Using the approximate integration approach of Chap. 4, derive an expression for the minimum time to climb in terms of $P_{S_{max}}$ at $n+1$ points along the climb path.

7.2 For the energy climb, derive the equation for the distance traveled while climbing from E_{S_0} to E_{S_f}. Assuming small flight path angle, show that the minimum distance climb is flown at the $V(E_S)$ where

$$\frac{\partial}{\partial V}\left(\frac{P_S}{V}\right)\bigg|_{E_S=\mathrm{Const}} = 0.$$

7.3 For the energy climb, derive the equation for the fuel consumed while climbing from E_{S_0} to E_{S_f}. Show that the minimum fuel climb is flown at the $V(E_S)$ where

$$\frac{\partial}{\partial V}\left(\frac{P_S}{CT}\right)\bigg|_{E_S=\mathrm{Const}} = 0.$$

7.4 Find the velocity profile $V(E_S)$ that maximizes the distance in nonsteady gliding flight $(T = 0)$ with γ small and $\dot{\gamma}$ negligible. Assume an Ideal Subsonic Airplane $(C_D = C_{D_0} + KC_L^2)$ and an exponential atmosphere $(\rho = \rho_s \exp(-h\lambda))$. First, show that the distance is given by

$$x_f - x_0 = \int_{E_{s_0}}^{E_{s_f}} \frac{V}{-P_S}\,dE_S = \int_{E_{s_0}}^{E_{s_f}} \frac{W}{D(E_S,V)}\,dE_S$$

so that maximum distance occurs when D is a minimum as in

$$\frac{\partial D}{\partial V}\bigg|_{E_S=\mathrm{Const}} = 0.$$

Next, show that

$$\frac{\partial D}{\partial V}\bigg|_{E_S=\mathrm{Const}} = \left(C_{D_0}S - \frac{KW^2}{\bar{q}^2 S}\right)\frac{\partial \bar{q}}{\partial V}\bigg|_{E_S=\mathrm{Const}}$$

and that, with $h = E_S - V^2/2g$,

$$\left.\frac{\partial \rho}{\partial V}\right|_{E_S=\text{Const}} = \rho V\left(\frac{V^2}{2\lambda g}+1\right) > 0.$$

Hence, minimum drag occurs when

$$C_{D_0}S - \frac{KW^2}{\bar{q}^2 S} = 0 \Rightarrow \bar{q} = \frac{W}{SC_L^*}$$

or when

$$(1/2)\rho_s e^{-\left(\frac{E_S-V^2/2g}{\lambda}\right)}V^2 = \frac{W}{SC_L^*}$$

which is to be solved for $V(E_S)$.

7.5 Consider the constant lift coefficient nonsteady glide of a low-speed airplane, $C_D = C_D(C_L)$, and assume that the flight path is shallow (γ small) and smooth ($\dot\gamma$ negligible). Show that

$$\begin{aligned} V &= \sqrt{\frac{2W}{\rho S C_L}} \\ x_f - x_0 &= E(C_L)(E_{S_0} - E_{S_f}) \end{aligned}$$

along the flight path. Hence, if the initial and final specific energies are prescribed maximum range occurs when the lift coefficient is that for maximum lift-to-drag ratio.

7.6 Consider the constant altitude deceleration of the Ideal Subsonic Airplane in gliding flight ($T = 0$) from initial velocity V_0 to final velocity V_f. Since the drag polar is parabolic with constant coefficients, show that the distance traveled is given by

$$x_f - x_0 = \frac{V^{*2}E^*}{2g}[F(u_0) - F(u_f)]$$

where $u = V/V^*$ and where

$$F(u) = \ln(1 + u^4).$$

7.7 Consider the constant altitude acceleration of the Ideal Subsonic Airplane at constant power setting from initial velocity V_0 to final velocity V_f. Assuming that the weight is constant, show that the distance traveled is given by

$$x_f - x_0 = \frac{V^{*2} E^*}{g} [A(u_0) - A(u_f)]$$

where

$$A(u) = \frac{1}{u_1^2 - u_2^2} \left[u_1^2 \ln(u_1^2 - u^2) - u_2^2 \ln(u^2 - u_2^2) \right]$$

where

$$u_1 = \sqrt{\tau + \sqrt{\tau^2 - 1}} , \quad u_2 = \sqrt{\tau - \sqrt{\tau^2 - 1}}$$

and where $\tau = T/(W/E^*)$.

7.8 A constant velocity turn is characterized by a constant turn rate $\dot{\psi}$ so that $\psi = \dot{\psi} t$. Integrate the equations of motion (7.56) and (7.57) to show that the trajectory is the circle

$$x^2 + [y - (V/\dot{\psi})]^2 = (V/\dot{\psi})^2$$

where $V/\dot{\psi}$ is the radius. Sketch the turn in the xy plane.

7.9 For a turn of the ISA, show that the load factor has the maximum value

$$n = \tau$$

which occurs when

$$u = \tau^{1/2} .$$

7.10 For a turn of an ISA, show that the bank angle has the maximum value

$$\mu = \arccos(1/\tau)$$

when

$$u = \tau^{1/2}.$$

7.11 NASA uses a large jet aircraft, affectionately called the Vomit
Comet, to give people the experience of "weightlessness" or "zero
g" without having to put them in orbit. Actually, astronauts in
orbit are not "weightless", nor are they in "zero g." In earth orbit,
g is roughly the same as it is on earth. The proper name for this
state is "free fall." From the equations of motion for nonsteady
flight over a flat earth, Sec. 2.5, it is seen that the airplane has
two controls, the angle of attack and the power setting. How should
the airplane be flown so that its trajectory is part of an orbit. How
should the trajectory be started to get the highest amount of free
fall time.

Ans. The airplane should be flown such that $T\cos\varepsilon - D = 0$ and
$T\sin\varepsilon + L = 0$. The initial flight path angle should be as high as
possible.

Chapter 8

6DOF Model: Wind Axes

In Chap. 2, the translational equations have been uncoupled from the rotational equations by assuming that the aircraft is not rotating and that control surface deflections do not affect the aerodynamic forces. The scalar equations of motion for flight in a vertical plane have been derived in the wind axes system. These equations have been used to study aircraft trajectories (performance). If desired, the elevator deflection history required by the airplane to fly a particular trajectory can be obtained by using the rotational equation.

In this chapter, the *six-degree-of-freedom (6DOF) model* for nonsteady flight in a vertical plane is presented in the wind axes system. Formulas are derived for calculating the forces and moments. Because it is possible to do so, the effect of elevator deflection on the lift is included. These results will be used in the next chapter to compute the elevator deflection required for a given flight condition. Finally, since the equations for the aerodynamic pitching moment are now available, the formula for the drag polar can be improved by using the trimmed polar. The aerodynamics of this chapter is based on Ref. Ho, which for straight-tapered wings has been summarized in Refs. Ro1 and Ro2.

8.1 Equations of Motion

The translational equations for nonsteady flight in a vertical plane in the wind axes system are given by Eqs. (2.24). The coordinate systems, the angles, the forces, and the moment about the center of gravity are

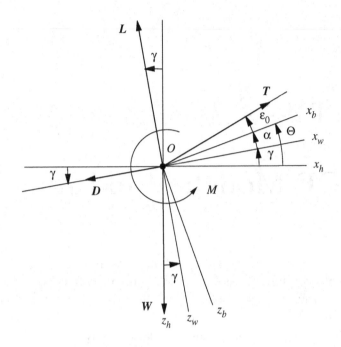

Figure 8.1: Forces and Moment

shown in Fig. 8.1 where Θ is the pitch angle and M is the resultant external moment (pitching moment). From dynamics, it is known that the pitching motion of an airplane is governed by the equation

$$M = I_{yy}\ddot{\Theta} \tag{8.1}$$

where I_{yy} is the *mass moment of inertia* and $\ddot{\Theta}$ is the *angular accelera-tion*, both about the y_w axis. If the *pitch rate* $Q = \dot{\Theta}$ is introduced, this second-order equation can be replaced by two first-order equations:

$$\begin{aligned} \dot{\Theta} &= Q \\ \dot{Q} &= M/I_{yy}. \end{aligned} \tag{8.2}$$

In general, the moments acting on an airplane are due to the thrust force and the aerodynamic force being moved to the center of gravity and due to the gyroscopic effects of the rotating masses in the engines (Ref. Mi). For the longitudinal motion of a conventional air-plane, the gyroscopic moment vector lies in the vertical plane. Hence, it causes small rotations about the roll and yaw axes (lateral-directional motion). For small perturbations, lateral-directional motion does not cause longitudinal motion. Therefore, the gyroscopic moment does not affect longitudinal motion so that the pitching moment is given by

$$M = M^T + M^A. \tag{8.3}$$

The complete set of the *6DOF equations of motion* for flight in a vertical plane in the wind axes system is given by

$$
\begin{aligned}
\dot{x} &= V \cos\gamma \\
\dot{h} &= V \sin\gamma \\
\dot{V} &= (g/W)[T\cos(\alpha + \varepsilon_0) - D - W\sin\gamma] \\
\dot{\gamma} &= (g/WV)[T\sin(\alpha + \varepsilon_0) + L - W\cos\gamma] \\
\dot{W} &= -CT \\
\dot{\Theta} &= Q \\
\dot{Q} &= M/I_{yy}
\end{aligned} \tag{8.4}
$$

where from Fig. 8.1

$$\Theta = \gamma + \alpha. \tag{8.5}$$

These equations contain several quantities (D, L, T, C) which are functions of variables already present.

If the airplane is pitching nose up about the center of gravity, the downward motion of the tail increases its angle of attack. This increases the lift of the horizontal tail and, in turn, opposes the rotational motion. This effect is modeled by including the pitch rate Q in the aerodynamics. There is a wing contribution from the wing, but it has been modeled in the literature by increasing the tail contribution by 10%.

It is assumed that the flow field around the airplane instantaneously adjusts itself to angle of attack and velocity changes. This is possible because these changes are not made rapidly. The deflection of the free stream by the wing, called downwash, moves from the wing to the horizontal tail in a finite time. This is modeled by assuming that the aerodynamics is a function of $\dot{\alpha}$.

The effects of Q and $\dot{\alpha}$ on the thrust, specific fuel consumption, drag, and thrust moment are neglected. If the effect of elevator deflection δ_E on the drag is also neglected, the propulsion and aerodynamic terms in the equations of motion satisfy the following functional relations:

$$T = T(h, V, P), \quad C = C(h, V, P)$$
$$D = D(h, V, \alpha), \quad L = L(h, V, \alpha, \delta_E, Q, \dot{\alpha}) \tag{8.6}$$
$$M^T = M^T(h, V, P), \quad M^A = M^A(h, V, \alpha, \delta_E, Q, \dot{\alpha}).$$

Eqs. (8.4) and (8.5) involve ten variables (eight states x, h, V, γ, W, Θ, Q, α and two controls P and δ_E). Hence, there are two mathematical degrees of freedom. To solve these equations, two additional equations involving existing variables must be provided. An example is specifying the two control histories $P = P(t)$ and $\delta_E(t)$, which are the controls available to the pilot, depending on the design of the horizontal tail.

Given the aerodynamics and propulsion characteristics of an airplane, these equations can be used to perform a numerical simulation of its pitching motion. One use of such a simulation is to study the stability characteristics of an airplane and its response to control and gust inputs throughout the flight envelope. Since the motion is only of interest for a short period of time, the atmospheric properties and the mass properties can be assumed constant. As a result, the kinematic equations and the mass equation uncouple from the system, and only the dynamic equations are relevant.

8.2   Aerodynamics and Propulsion

To derive formulas for predicting the aerodynamics, it is necessary to write the forces and moments in coefficient form. First, the lift and aerodynamic pitching moment satisfy the functional relations

$$\begin{aligned} C_L &= C_L(\alpha, \delta_E, M, \bar{c}Q/2V, \bar{c}\dot{\alpha}/2V) \\ C_m^A &= C_m^A(\alpha, \delta_E, M, \bar{c}Q/2V, \bar{c}\dot{\alpha}/2V). \end{aligned} \tag{8.7}$$

where M is the Mach number. In the derivations that follow, it is shown that C_L and C_m^A are linear in every variable except Mach number. Hence, C_L and C_m^A are written as

$$\begin{aligned} C_L &= C_{L_0}(M) + C_{L_\alpha}(M)\alpha + C_{L_{\delta_E}}(M)\delta_E \\ &+ C_{L_Q}(M)(\bar{c}Q/2V) + C_{L_{\dot{\alpha}}}(M)(\bar{c}\dot{\alpha}/2V) \end{aligned} \tag{8.8}$$

$$C_m^A = C_{m_0}^A(M) + C_{m_\alpha}^A(M)\alpha + C_{m_{\delta_E}}^A(M)\delta_E \qquad (8.9)$$
$$+ C_{m_Q}^A(M)(\bar{c}Q/2V) + C_{m_{\dot{\alpha}}}^A(M)(\bar{c}\dot{\alpha}/2V)$$

where C_{L_α}, for example, denotes the partial derivative of C_L with respect to α. Also, by definition

$$C_{L_Q} \triangleq \frac{\partial C_L}{\partial \frac{\bar{c}Q}{2V}}, \quad C_{m_Q}^A \triangleq \frac{\partial C_m^A}{\partial \frac{\bar{c}Q}{2V}}, \quad C_{L_{\dot{\alpha}}} \triangleq \frac{\partial C_L}{\partial \frac{\bar{c}\dot{\alpha}}{2V}}, \quad C_{m_{\dot{\alpha}}}^A \triangleq \frac{\partial C_m^A}{\partial \frac{\bar{c}\dot{\alpha}}{2V}}. \qquad (8.10)$$

For airplanes with moveable horizontal tails, it is possible to separate the i_H term from C_{L_0} and $C_{m_0}^A$. Then, i_H would appear as a variable along with δ_E.

The effects of δ_E, Q, and $\dot{\alpha}$ on the drag are negligible so that its functional relation is given by

$$C_D = C_D(\alpha, M). \qquad (8.11)$$

For a parabolic drag polar, this functional relation has the form

$$C_D = C_0(M) + C_1(M)\alpha + C_2(M)\alpha^2. \qquad (8.12)$$

Finally, the nondimensional quantities C_T and C_m^T are not really force and moment coefficients in the sense that they are not functions of other nondimensional quantities. These quantities satisfy functional relations of the form

$$C_T = C_T(h, M, P), \quad C_m^T = C_m^T(h, M, P). \qquad (8.13)$$

In the remainder of this chapter, formulas are developed for the Mach number dependent terms. However, because the derivation of the pitching moment follows a different approach than the derivation of the lift and drag for trajectory analysis, it is necessary to start over. First, airfoils, wings, wing-body combinations, downwash at the horizontal tail, and control surfaces are discussed. Next, formulas are derived for airplane lift and airplane pitching moment for quasi-steady flight ($Q = \dot{\alpha} = 0$). Subsequently, formulas are developed for the Q and $\dot{\alpha}$ terms (nonsteady flight). Finally, airplane drag is discussed, as is the trimmed drag polar.

8.3 Airfoils

An airfoil (two-dimensional wing) at an angle of attack experiences a resultant aerodynamic force, and the point on the chord through which the line of action passes is called the center of pressure (Fig. 3.6). The resultant aerodynamic force can be resolved into components parallel and perpendicular to the velocity vector, that is, the drag and the lift. Unfortunately, the center of pressure varies with V and α. The forces can be replaced by forces and a moment at a fixed point. In trying to select the best fixed point, it has been discovered that a point on the chord exists where the moment is independent of α. This point is called the *aerodynamic center*, and it is located approximately at the quarter chord at subsonic speeds. Hence, the procedure for representing the resultant aerodynamic force becomes that of Fig. 8.2.

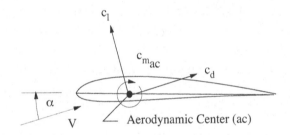

Figure 8.2: Airfoil Aerodynamic Center

In general,

$$\alpha = \alpha(c_l, M), \quad c_d = c_d(c_l, M), \quad c_{m_{ac}} = c_{m_{ac}}(M) \qquad (8.14)$$

where M is the Mach number. By holding M constant and varying the lift coefficient, the above quantities vary as in Fig. 8.3. Away from the maximum lift coefficient, it is seen that angle of attack varies linearly with the lift coefficient, the drag coefficient varies quadratically with c_l, and the pitching moment about the aerodynamic center is independent of c_l.

A set of data for the NACA 64-109 airfoil (Ref. AD or Ho) is presented in Table 8.1. While some of the numbers are presented in terms of degrees, the use of the numbers in all formulas is in radians. Only the lift-curve slope varies with Mach number at subsonic speeds.

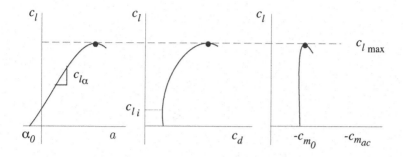

Figure 8.3: Airfoil Aerodynamic Characteristics (M given)

Table 8.1: Data for the NACA 64-109 Airfoil

Parameter	M = 0	Variation with Mach number
α_0	-0.5 deg	negligible
c_{l_α}	0.110 deg^{-1}	$c_{l_\alpha} = \frac{(c_{l_\alpha})_{M=0}}{\sqrt{1-M^2}}$
ac/c	0.258	negligible
$c_{m_{ac}}$	-0.0175	negligible

8.4   Wings and Horizontal Tails

The aerodynamic characteristics of a three-dimensional wing are shown in Fig. 8.4. In stability and control studies, the drag plays a small role relative to the lift. Hence, the aerodynamic parameters of a wing are represented by the aerodynamic center, the lift coefficient, and the pitching moment coefficient about the aerodynamic center.

The *aerodynamic center* of a half wing (Fig. 8.5) is defined in terms of an *equivalent rectangular wing*, that is, a rectangular wing which has the same planform area, the same lift, the same moment about the y axis, and the same moment about the x axis. Then, the aerodynamic center of the equivalent rectangular wing is the aerodynamic center of the original wing, and the chord of the equivalent rectangular wing is the *mean aerodynamic chord*. For an straight-tapered wing, the location of the aerodynamic center ξ, η and the mean aerodynamic chord $\bar{c}$ are

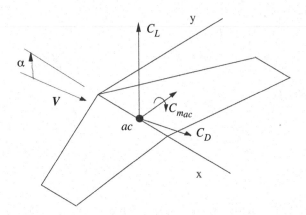

Figure 8.4: Aerodynamic Parameters of a 3D Wing

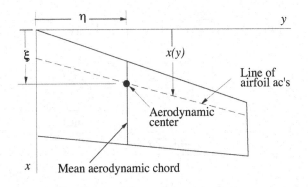

Figure 8.5: Aerodynamic Center and Mean Aerodynamic Chord

given by

$$\xi = p\bar{c} + \eta \tan \Lambda_{le}, \quad \eta = \frac{b}{6}\frac{1+2\lambda}{1+\lambda}, \quad \bar{c} = \frac{2c_r}{3}\frac{1+\lambda+\lambda^2}{1+\lambda} \qquad (8.15)$$

where p is the location of the airfoil ac as a fraction of the chord ($p \cong .25$) and Λ_{le} is the leading edge sweep. These results are for a half wing. For a whole wing, the aerodynamic center and the mean aerodynamic chord are located on the x axis. The aerodynamic center does not change location with Mach number at subsonic speeds. At transonic speeds, it transitions from the subsonic quarter-chord location to the supersonic half-chord location.

There is a simple graphical procedure for finding ξ, η, and $\bar{c}$ as shown in Fig. 8.6. The algorithm is as follows:

1. Add the tip chord behind the root chord to locate point A.

2. Add the root chord in front of the tip chord to locate point B.

3. Connect points A and B with a straight line.

4. Locate the half-root-chord point C and the half-tip-chord point D, and connect them with a straight line.

5. Where the lines AB and CD intersect, that is, at E, draw the chord. This is the mean aerodynamic chord.

6. Where the mean aerodynamic chord and the line of airfoil aerodynamic centers intersect is the location of the wing aerodynamic center.

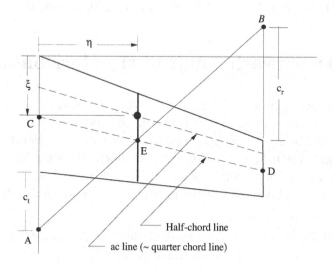

Figure 8.6: Graphical Procedure

The lift coefficient $C_L = C_{L_\alpha}(\alpha - \alpha_{0L})$ is determined by α_{0L} and C_{L_α}, which are given by (Sec. 3.5)

$$\alpha_{0L} = \alpha_0 \qquad (8.16)$$

and

$$C_{L_\alpha} = \frac{\pi A}{1 + \sqrt{1 + (A/2\kappa)^2[1 + \tan^2 \Lambda_{hc} - M^2]}}. \tag{8.17}$$

The zero-lift angle of attack does not change with Mach number, and the lift-curve slope increases.

The airfoil pitching moment about the aerodynamic center does not vary with the angle of attack. For a wing with the same airfoil shapes, $c_{m_{ac}}$ does not vary along the span so that the wing value equals the airfoil value. Hence,

$$C_{m_{ac}} = c_{m_{ac}} . \tag{8.18}$$

At subsonic speeds, $C_{m_{ac}}$ does not vary with the Mach number.

The addition of the body to the wing to form the *wing-body combination* has three effects: the ac moves forward, C_{L_α} increases, and $C_{m_{ac}}$ decreases. In general, these effects are small and can be omitted in a first calculation. In other words, the aerodynamic characteristics of the wing-body combination can be taken to be those of the entire wing alone (Ref. Ne, Fig. 2.17, p. 5.8, Ref. Pa, p. 239, or Ref. Ro2, p. 3.45)

8.5   Downwash Angle at the Horizontal Tail

The tail is in the flow field behind the wing. To produce lift, the wing must deflect the free stream downward. In addition, the wing slows that part of the air stream which passes near the wing. Hence, the tail operates in a flow field which is deflected through an angle ε from the free stream and which is moving at a slightly lower speed. The downwash at the horizontal tail depends on the location of the horizontal tail relative to the wing.

To calculate the *downwash angle* at the horizontal tail (ε), it is assumed to vary linearly with the angle of attack as

$$\varepsilon = \varepsilon_\alpha(\alpha - \alpha_{0L}) \tag{8.19}$$

The slope ε_α of this line behaves like the lift-curve slope C_{L_α} and is assumed to satisfy the relation

$$\varepsilon_\alpha = (\varepsilon_\alpha)_{M=0} \frac{C_{L_\alpha}}{(C_{L_\alpha})_{M=0}}. \tag{8.20}$$

In turn, the zero-Mach-number value is approximated by

$$(\varepsilon_\alpha)_{M=0} = 4.44(K_A K_\lambda K_H \sqrt{\cos \Lambda_{qc}} \,)^{1.19} \tag{8.21}$$

where Λ_{qc} is the quarter-chord sweep angle and where

$$K_A = \frac{1}{A} - \frac{1}{1 + A^{1.7}}, \quad K_\lambda = \frac{10 - 3\lambda}{7}, \quad K_H = \frac{1 - \frac{h_H}{b}}{\sqrt[3]{\frac{2l_H}{b}}}. \tag{8.22}$$

The quantities l_H and h_H locate the horizontal tail relative to the wing as shown in Fig. 8.7. The quantity l_H is the distance along the mean

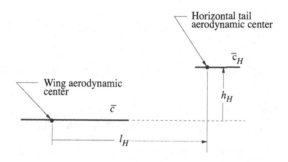

Figure 8.7: Location of the Horizontal Tail

aerodynamic chord of the wing from the wing aerodynamic center to the horizontal tail aerodynamic center. Also, h_H is the distance of the tail aerodynamic center above the line of $\bar{c}$.

The decrease in the speed of the air stream behind the wing is taken into account by introducing the *tail efficiency factor*

$$\eta_H = \frac{\bar{q}_H}{\bar{q}} = 0.9. \tag{8.23}$$

In this formula, $\bar{q}_H$ is the dynamic pressure in front of the horizontal tail and $\bar{q}$ is the dynamic pressure in front of the wing (free stream dynamic pressure). In an incompressible flow, this assumption represents a 5% change in the velocity from in front of the wing to in front of the horizontal tail.

8.6   Control Surfaces

The control for the pitching motion of a conventional airplane is the
elevator which is the aft portion of the horizontal tail. A positive elevator
deflection is trailing edge down and leads to a negative pitching moment.

A cambered airfoil with a control flap is shown in Fig. 8.8 with
a positive flap deflection. The aerodynamics of such a wing section are
presented schematically in Fig. 8.9. Note that c_l varies linearly with α
for a nonzero δ, that α_0 decreases as δ increases, that c_{l_α} does not change,
the maximum lift coefficient increases, and that $c_{m_{ac}}$ decreases. The
location of the aerodynamic center does not change with flap deflection.

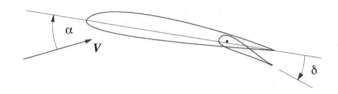

Figure 8.8: Airfoil with Deflected Flap

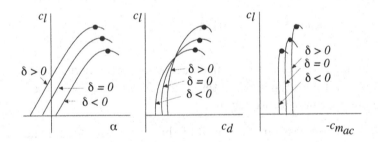

Figure 8.9: Effects of Flap on Airfoil Aerodynamics (M given)

For moderate control deflections, the lift coefficient varies linearly with δ. Hence, it is assumed that

$$c_l = c_{l_0} + c_{l_\alpha}\alpha + c_{l_\delta}\delta \tag{8.24}$$

where c_{l_0}, c_{l_α} and c_{l_δ} are known for a particular airfoil. For horizontal
tails that must produce downward lift as effectively as upward lift, the

airfoils are symmetric so that c_{l_0} is zero. Next, Eq. (8.21) is rewritten in the form

$$c_l = c_{l_0} + c_{l_\alpha}(\alpha + \bar{\tau}\delta) \qquad (8.25)$$

where $\bar{\tau} = c_{l_\delta}/c_{l_\alpha}$ is called the *airfoil control effectiveness* and is plotted in Fig. 8.10 as a function of the flap size. The control effectiveness does

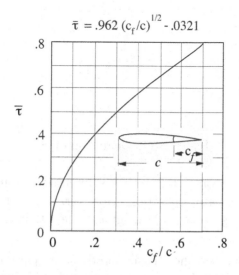

Figure 8.10: Airfoil Control Effectiveness

not change with Mach number at subsonic speeds.

For a wing with a control surface, a positive control surface deflection leaves the aerodynamic center unchanged, decreases α_{0L}, does not change C_{L_α}, and decreases $C_{m_{ac}}$. For moderate control surface deflections, the lift is linear in δ and is approximated by

$$C_L = C_{L_o} + C_{L_\alpha}(\alpha + \tau\delta) \qquad (8.26)$$

The *control surface effectiveness* τ is given by

$$\tau = \frac{2}{S} \int_0^{\frac{b}{2}} \bar{\tau}(y)c(y)\,dy \qquad (8.27)$$

where $\bar{\tau}(y)$ is assumed to be constant over that part of the wing which has the control surface and zero over the remainder of the wing. Hence,

$$\tau = \frac{S_c}{S}\bar{\tau} \qquad (8.28)$$

where S_c is the planform area of the wing associated with the control surface (see Fig. 8.11). For a horizontal tail with a full-span elevator, $S_c = S$ so that $\tau = \bar{\tau}$.

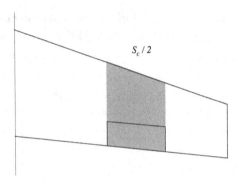

Figure 8.11: Definition of S_c

For a horizontal tail with a symmetric airfoil, C_{mac_H} is proportional to the elevator deflection. However, its effect on the airplane pitching moment is an order of magnitude less that of C_{mac_W}, so it is not discussed.

A summary of how basic aerodynamic parameters vary with control deflection and Mach number is presented in Table 8.2.

Table 8.2: Changes with Control Deflection and Mach Number

Subsonic airfoil	Subsonic wing	Changes with δ	Changes with M
ac	ac	no	no
α_0	α_{0L}	yes	no
c_{l_α}	C_{L_α}	no	yes
c_{mac}	C_{mac}	yes	no
$\bar{\tau}$	τ	no	no

8.7   Airplane Lift

The coordinate system and other information for computing the lift and the aerodynamic pitching moment of an airplane are shown in Fig. 8.12.

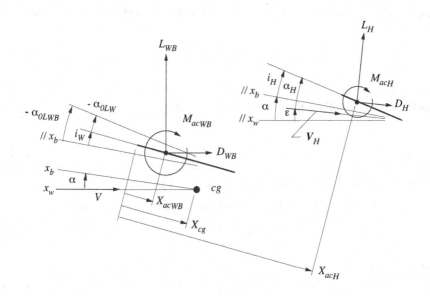

Figure 8.12: Coordinate System for Forces and Moments

The wing-body configuration is represented by its forces and moment acting at the aerodynamic center of the wing mean aerodynamic chord, which is at an incidence i_W with respect to the x_b axis. It is recalled that in all calculations, wing-body quantities are replaced by the corresponding wing quantities. The horizontal tail is represented by its forces and moment acting at the aerodynamic center of the horizontal tail mean aerodynamic chord. It has an *incidence i_H* relative to the x_b axis. The angle of attack of the airplane (wing-body combination) is the angle between the x_b axis and the x_w axis which is along the velocity vector. For the horizontal tail, the angle of attack is seen to be $\alpha_H = \alpha + i_H - \varepsilon$. The coordinate X which locates the horizontal tail is measured from the leading edge of the wing mean aerodynamic chord.

In order to derive the equation for C_L, consider the components of the airplane. If small angles are assumed and contributions of the drags are neglected, the lift of the airplane is given by

$$L = L_{WB} + L_H. \tag{8.29}$$

In order to have the horizontal tail behave as a wing in a free stream, its lift coefficient is formed by dividing by the dynamic pressure at the horizontal tail and by the planform area of the horizontal tail, that is,

$$C_{L_H} = \frac{L_H}{\bar{q}_H S_H} . \tag{8.30}$$

As a consequence, if Eq. (8.29) is divided by $\bar{q}S$ to form the airplane lift coefficient, the following result is obtained:

$$C_L = C_{L_{WB}} + C_{L_H} \eta_H \frac{S_H}{S} \tag{8.31}$$

where the tail efficiency factor η_H is defined in Eq. (8.23).

The wing-body lift coefficient can be expressed as

$$C_{L_{WB}} = C_{L_{\alpha_{WB}}} (\alpha_{WB} - \alpha_{0L_{WB}}) \tag{8.32}$$

where

$$\alpha_{WB} \overset{\Delta}{=} \alpha \tag{8.33}$$

and

$$\alpha_{0L_{WB}} = -(i_W - \alpha_{0L_W}) \tag{8.34}$$

from Fig. 8.12. For a straight-tapered wing with the same airfoil shapes and no twist, $\alpha_{0L_W} = \alpha_0$ which is the zero-lift angle of attack of the airfoil. The combination of Eqs. (8.32) through (8.34) leads to

$$C_{L_{WB}} = C_{L_{0_{WB}}} + C_{L_{\alpha_{WB}}} \alpha \tag{8.35}$$

where

$$C_{L_{0_{WB}}} = C_{L_{\alpha_{WB}}} (i_W - \alpha_{0L_W}) \tag{8.36}$$

and

$$C_{L_{\alpha_{WB}}} = C_{L_{\alpha_W}}, \quad \alpha_{0L_W} = \alpha_0 \tag{8.37}$$

because the aerodynamic characteristics of the wing-body combination are being approximated by those of the wing alone.

Next, from Eq. (8.26), the horizontal tail lift coefficient can be written as

$$C_{L_H} = C_{L_{0_H}} + C_{L_{\alpha_H}} (\alpha_H + \tau_E \delta_E). \tag{8.38}$$

Since a horizontal tail should be effective in producing up and down lift, its airfoil shape is symmetric, and $C_{L_{0_H}}$ is zero. From Fig. 8.11, it is seen that the angle of attack of the horizontal tail is given by

$$\alpha_H = \alpha + i_H - \varepsilon. \tag{8.39}$$

The downwash angle at the horizontal tail is given by Eq. (8.19) to be

$$\varepsilon = \varepsilon_\alpha(\alpha_{WB} - \alpha_{0LWB}) = \varepsilon_\alpha(\alpha + i_W - \alpha_{0LW}) = \varepsilon_0 + \varepsilon_\alpha\alpha \qquad (8.40)$$

where

$$\varepsilon_0 = \varepsilon_\alpha(i_W - \alpha_{0LW}). \qquad (8.41)$$

Finally, the lift coefficient of the horizontal tail becomes

$$C_{L_H} = C_{L_{\alpha_H}}[i_H - \varepsilon_0 + (1 - \varepsilon_\alpha)\alpha + \tau_E\delta_E]. \qquad (8.42)$$

For the airplane, the combination of Eqs. (8.31), (8.35), and (8.42) leads to

$$
\begin{aligned}
C_L &= \left[C_{L_{\alpha_{WB}}}(i_W - \alpha_{0LW}) + C_{L_{\alpha_H}}(i_H - \varepsilon_0)\eta_H\tfrac{S_H}{S}\right] \\
&+ \left[C_{L_{\alpha_{WB}}} + C_{L_{\alpha_H}}(1 - \varepsilon_\alpha)\eta_H\tfrac{S_H}{S}\right]\alpha \qquad (8.43) \\
&+ \left[C_{L_{\alpha_H}}\tau_E\eta_H\tfrac{S_H}{S}\right]\delta_E
\end{aligned}
$$

The form of Eq. (8.43) is

$$C_L = C_{L_0}(M) + C_{L_\alpha}(M)\alpha + C_{L_{\delta_E}}(M)\delta_E, \qquad (8.44)$$

where

$$C_{L_0}(M) = C_{L_{\alpha_{WB}}}(i_W - \alpha_{0LW}) + C_{L_{\alpha_H}}(i_H - \varepsilon_0)\eta_H\frac{S_H}{S} \qquad (8.45)$$

$$C_{L_\alpha}(M) = C_{L_{\alpha_{WB}}} + C_{L_{\alpha_H}}(1 - \varepsilon_\alpha)\eta_H\frac{S_H}{S} \qquad (8.46)$$

$$C_{L_{\delta_E}}(M) = C_{L_{\alpha_H}}\tau_E\eta_H\frac{S_H}{S} \qquad (8.47)$$

For the SBJ (App. A) at $M = 0.6$, straightforward calculations lead to $C_{L_0} = .0835$, $C_{L_\alpha} = 5.16$, and $C_{L_{\delta_E}} = .430$.

8.8 Airplane Pitching Moment

The airplane pitching moment is the sum of the *thrust pitching moment* and the *aerodynamic pitching moment*. The aerodynamic pitching moment is discussed first, followed by the thrust pitching moment and the airplane pitching moment.

8.8.1   Aerodynamic pitching moment

In Fig. 8.12, moment arms are measured in a coordinate system whose origin is at the leading edge of the wing mean aerodynamic chord. Because i_W is very small, the distance X can be measured along the mean aerodynamic chord or along the x_b axis. In either case, it is positive toward the tail.

The moments due to the drags of the wing-body combination and the horizontal tail are neglected because they are small relative to the moments of the lifts. Also, the horizontal tail moment about the aerodynamic center due to δ_E is neglected. After these assumptions are taken into account, the aerodynamic pitching moment about the center of gravity becomes

$$M^A = L_{WB}(X_{cg} - X_{ac_{WB}}) + M_{ac_{WB}} - L_H(X_{ac_H} - X_{cg}). \tag{8.48}$$

In coefficient form, that is, after dividing through by $\bar{q}S\bar{c}$ and accounting for Eq. (8.30), this equation becomes

$$C_m^A = C_{L_{WB}}(\bar{X}_{cg} - \bar{X}_{ac_{WB}}) + C_{m_{ac_{WB}}} - C_{L_H}\eta_H\frac{S_H}{S}(\bar{X}_{ac_H} - \bar{X}_{cg}) \tag{8.49}$$

where $\bar{X}$ is defined as

$$\bar{X} = \frac{X}{\bar{c}}. \tag{8.50}$$

Eq. (8.49) in combination with the equations for $C_{L_{WB}}$ and C_{L_H} leads to the following expression for C_m^A:

$$\begin{aligned}
C_m^A &= \left[C_{L_{\alpha_{WB}}}(i_W - \alpha_{0L_W})(\bar{X}_{cg} - \bar{X}_{ac_{WB}}) + C_{m_{ac_{WB}}} \right.\\
&\quad \left. - C_{L_{\alpha_H}}(i_H - \varepsilon_0)\eta_H\bar{V}_H\right] \\
&\quad + \left[C_{L_{\alpha_{WB}}}(\bar{X}_{cg} - \bar{X}_{ac_{WB}}) - C_{L_{\alpha_H}}(1 - \varepsilon_\alpha)\eta_H\bar{V}_H\right]\alpha \\
&\quad + \left[-C_{L_{\alpha_H}}\tau_E\eta_H\bar{V}_H\right]\delta_E
\end{aligned} \tag{8.51}$$

Here, $\bar{V}_H$ is the horizontal tail volume coefficient which is defined as

$$\bar{V}_H = \frac{S_H}{S}(\bar{X}_{ac_H} - \bar{X}_{cg}). \tag{8.52}$$

and varies with cg location.

The form of Eq. (8.51) is

$$C_m^A = C_{m_0}^A(M) + C_{m_\alpha}^A(M)\alpha + C_{m_{\delta_E}}^A(M)\delta_E \tag{8.53}$$

where

$$C_{m_0}^A(M) = C_{L_{\alpha_{WB}}}(i_W - \alpha_{0LW})(\bar{X}_{cg} - \bar{X}_{ac_{WB}}) + C_{m_{ac_{WB}}}$$
$$- C_{L_{\alpha_H}}(i_H - \varepsilon_0)\eta_H \bar{V}_H \qquad (8.54)$$
$$C_{m_\alpha}^A(M) = C_{L_{\alpha_{WB}}}(\bar{X}_{cg} - \bar{X}_{ac_{WB}}) - C_{L_{\alpha_H}}(1 - \varepsilon_\alpha)\eta_H \bar{V}_H \quad (8.55)$$
$$C_{m_{\delta_E}}^A(M) = -C_{L_{\alpha_H}}\tau_E \eta_H \bar{V}_H. \qquad (8.56)$$

While not explicitly shown, all of these quantities is also a function of the cg position.

For the SBJ at $M = 0.6$ and $\bar{X}_{cg} = .30$, straightforward calculations lead to $\bar{V}_H = .612$, $C_{m_0}^A = .0895$, $C_{m_\alpha}^A = -1.09$, and $C_{m_{\delta_E}}^A = -1.13$.

In general, the aerodynamic center is defined as that point about which the pitching moment is independent of the angle of attack. If the center of gravity is imagined to be an arbitrary reference point, then C_m is independent of α if $\bar{X}_{cg}$ is such that the coefficient of α, that is, $C_{m_\alpha}^A$, is zero. This cg position is then the *aerodynamic center* of the airplane and is denoted by $\bar{X}_{ac}$. Hence, if $\bar{X}_{cg}$ in Eq. (8.55) is replaced by $\bar{X}_{ac}$ (recalling the definition (8.52) of $\bar{V}_H$) and $C_{m_\alpha}^A$ is set equal to zero, the result can be solved for $\bar{X}_{ac}$ as follows

$$\bar{X}_{ac} = \frac{C_{L_{\alpha_{WB}}}\bar{X}_{ac_{WB}} + C_{L_{\alpha_H}}(1 - \varepsilon_\alpha)\eta_H \frac{S_H}{S}\bar{X}_{ac_H}}{C_{L_{\alpha_{WB}}} + C_{L_{\alpha_H}}(1 - \varepsilon_\alpha)\eta_H \frac{S_H}{S}}. \qquad (8.57)$$

The location of the aerodynamic center does not change with Mach number.

For the SBJ at $M = 0.6$, it is seen that $\bar{X}_{ac_{WB}} = 0.258$, $\bar{X}_{ac_H} = 2.93$, and $\bar{X}_{ac} = 0.512$.

For an arbitrary center of gravity location ($\bar{X}_{cg} \neq \bar{X}_{ac}$), the expression for $C_{m_\alpha}^A$ can be rewritten as

$$C_{m_\alpha}^A = C_{L_\alpha}(\bar{X}_{cg} - \bar{X}_{ac}). \qquad (8.58)$$

Since the lift-curve slope is positive. the sign of $C_{m_\alpha}^A$ depends on the location of the center of gravity relative to the aerodynamic center.

8.8.2 Thrust pitching moment

Jet engines on an airplane have a number of effects on the pitching moment. First, there is a moment due to the line of action of the thrust

force not passing through the center of gravity. Second, engines are usually at an angle of attack relative to the free stream so that the air stream
passing through the engine is deflected downward. This flow deflection
causes a force which acts in the neighborhood of the engine inlet. While
this force has a negligible effect on the translational motion of airplanes,
its corresponding moment also has a small effect on the rotational motion. Third, the high-speed jet stream exiting the engine pulls slower
moving air around it along with it (entrainment) and modifies the flow
field around the jet stream. To prevent this effect from changing the
flow field around the horizontal tail, the placement of the horizontal tail
is dictated by the location of the engines. For engines mounted under
the wings, the horizontal tail can be placed at the bottom of the vertical
tail. For fuselage-mounted engines, the horizontal tail is placed at the
top of the vertical tail.

The *thrust pitching moment* is written in the form

$$C_m^T = C_{m_0}^T + C_{m_\alpha}^T \alpha. \tag{8.59}$$

The term $C_{m_0}^T$ is the *direct thrust moment coefficient* (see Fig. 8.13)

$$C_{m_0}^T = \frac{T l_T}{\bar{q} S \bar{c}}. \tag{8.60}$$

where l_T is the moment arm of the thrust, positive when the thrust acts
below the center of gravity. For body-mounted engines, the moment
arm is sufficiently small that $C_{m_0}^T$ can be negligible with respect to the
corresponding aerodynamic moment term.

The term $C_{m_\alpha}^T \alpha$ represents the thrust moment coefficient due
to turning the inlet jet stream into the engine. For body-mounted engines, this term is insignificant because the wing turns the flow into the
engine. Wing-mounted engines operate in the upwash in front of the
wing which increases the angle of attack of the engine. Hence, the inlets
of these engines are placed out in front of the wing as far as possible.
Regardless, for low angle of attack operation, this term is not considered
to be important and is dropped from this point. Perhaps it is important
at high angle of attack and high power setting.

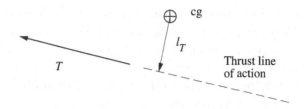

Figure 8.13: Direct Thrust Moment

8.8.3   Airplane pitching moment

The *airplane pitching moment* is the sum of the aerodynamic pitching moment (8.53) and the thrust pitching moment (8.59) and is given by

$$C_m = C_{m_0} + C_{m_\alpha}\alpha + C_{m_{\delta_E}}\delta_E. \qquad (8.61)$$

where

$$C_{m_0} = C_{m_0}^A + C_{m_0}^T, \quad C_{m_\alpha} = C_{m_\alpha}^A, \quad C_{m_{\delta_E}} = C_{m_{\delta_E}}^A. \qquad (8.62)$$

8.9   *Q* Terms

When an airplane is pitching up about the cg, the horizontal tail experiences an increase in its angle of attack due to the rotational motion as shown in Fig. 8.14. If the pitch rate is zero, air moves past the hori-

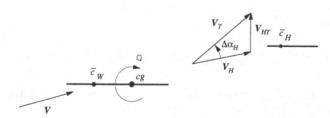

Figure 8.14: Increase in α_H due to Q

zontal tail at the speed V_H. If the pitch rate is not zero, the tail moves downward at the speed

$$V_{HT} = (X_{ac_H} - X_{cg})Q. \qquad (8.63)$$

This is the same as the air moving upward over the horizontal tail at the same speed. Consequently, the actual speed of the air over the horizontal tail is vector sum of V_H and V_{HT} or V_T as shown in the figure. The angle $\Delta\alpha_H$ between the velocity V_T and the velocity V_H is the increase in the angle of attack of the horizontal tail due to the rotational motion. If small angles and $V_H \cong V$ are assumed,

$$\Delta\alpha_H = \frac{V_{HT}}{V_H} = \frac{(X_{ac_H} - X_{cg})Q}{V} \tag{8.64}$$

so that α_H becomes

$$\alpha_H = \alpha + i_H - \epsilon + \Delta\alpha_H. \tag{8.65}$$

If Eq. (8.65) is used in place of Eq. (8.39) in the development of the lift coefficient, the lift coefficient becomes

$$C_L = (C_L)_{Q=0} + C_{L_{\alpha_H}}\eta_H\frac{S_H}{S}\frac{(X_{ac_H} - X_{cg})Q}{V}. \tag{8.66}$$

Then, the definition (8.10) of C_{L_Q} leads to

$$C_{L_Q} = 2C_{L_{\alpha_H}}\eta_H\bar{V}_H \tag{8.67}$$

where $\bar{V}_H$ is the horizontal tail volume coefficient. In the same vein, the pitching moment can be written as

$$C_m^A = (C_m^A)_{Q=0} - C_{L_{\alpha_H}}\eta_H\bar{V}_H\frac{X_{ac_H} - X_{cg}}{V}Q \tag{8.68}$$

so that the definition (8.10) of $C_{m_Q}^A$ gives

$$C_{m_Q}^A = -2C_{L_{\alpha_H}}\eta_H\bar{V}_H(\bar{X}_{ac_H} - \bar{X}_{cg}) = -C_{L_Q}(\bar{X}_{ac_H} - \bar{X}_{cg}). \tag{8.69}$$

Values of these quantities for the SBJ at $M = .6$ (App. A) are given by $C_{L_Q} = 4.44$ and $C_{m_Q}^A = $ -11.7.

8.10 $\dot{\alpha}$ Terms

The $\dot{\alpha}$ terms are included as an attempt to model the time dependence of the downwash at the horizontal tail. Imagine an airplane in gliding level flight where α is increasing with time. This means that the downwash angle at the horizontal tail ε is increasing with time. However, it takes a finite time for the downwash to get from the wing to the horizontal tail. Hence, at a given time instant, the downwash angle ($\dot\alpha \neq 0$) at the horizontal tail is the downwash angle ($\dot\alpha = 0$) at the wing at an earlier time, that is,

$$\varepsilon_{\dot\alpha \neq 0}(t) = \varepsilon(t - \Delta t) \tag{8.70}$$

where Δt is approximated by

$$\Delta t = \frac{X_{ac_H} - X_{ac_{WB}}}{V}. \tag{8.71}$$

For small Δt, a Taylor series expansion yields

$$\varepsilon_{\dot\alpha \neq 0}(t) = \varepsilon(t) - \dot\varepsilon(t)\Delta t. \tag{8.72}$$

Since $\varepsilon = \varepsilon_0 + \varepsilon_\alpha \alpha$, it is seen that

$$\varepsilon_{\dot\alpha \neq 0}(t) = \varepsilon(t) - \varepsilon_\alpha \dot\alpha \Delta t \tag{8.73}$$

Finally, the angle of attack of the horizontal tail is given by

$$\alpha_H = \alpha + i_H - \varepsilon + \varepsilon_\alpha \dot\alpha \frac{X_{ac_H} - X_{ac_{WB}}}{V}. \tag{8.74}$$

With regard to the lift coefficient, the use of Eq. (8.74) in place of Eq. (8.39) in the development of the lift coefficient leads to

$$C_L = (C_L)_{\dot\alpha = 0} + C_{L_{\alpha_H}} \eta_H \frac{S_H}{S} \varepsilon_\alpha \frac{X_{ac_H} - X_{ac_{WB}}}{V} \dot\alpha. \tag{8.75}$$

Hence, the definition (8.10) of $C_{L_{\dot\alpha}}$ gives

$$C_{L_{\dot\alpha}} = 2 C_{L_{\alpha_H}} \eta_H \frac{S_H}{S} \varepsilon_\alpha (\bar{X}_{ac_H} - \bar{X}_{ac_{WB}}). \tag{8.76}$$

A similar development for the pitching moment gives

$$C_m^A = (C_m^A)_{\dot\alpha = 0} - C_{L_{\alpha_H}} \eta_H \varepsilon_\alpha \frac{X_{ac_H} - X_{ac_{WB}}}{V} \bar{V}_H \dot\alpha \tag{8.77}$$

so that, the definition (8.10) of $C_{m_{\dot\alpha}}$ yields

$$C_{m_{\dot\alpha}}^A = -2\eta_H C_{L_{\alpha_H}} \varepsilon_\alpha (\bar{X}_{ac_H} - \bar{X}_{ac_{WB}}) \bar{V}_H = -C_{L_{\dot\alpha}}(\bar{X}_{ac_H} - \bar{X}_{cg}). \tag{8.78}$$

Values of these quantities for the SBJ at $M = .6$ (App. A) are given by $C_{L_{adot}} = 1.89$ and $C_{m_{adot}}^A = -4.98$.

8.11 Airplane Drag

The parabolic drag polar of a complete aircraft has been discussed in Chap. 3 and has the form

$$C_D = \bar{C}_{D_0} + \bar{K}C_L{}^2 \tag{8.79}$$

where $\bar{C}_{D_0}$ is the zero-lift drag coefficient and $\bar{K}$ is the induced drag factor. For subsonic speeds, both $\bar{C}_{D_0}$ and $\bar{K}$ are constant. When the Eq. (8.44) for C_L is substituted into the drag polar and terms involving δ_E and i_H are neglected, the equation for the *drag* becomes

$$C_D = C_0(M) + C_1(M)\alpha + C_2(M)\alpha^2. \tag{8.80}$$

where

$$
\begin{aligned}
C_0 &= \bar{C}_{D_0} + \bar{K}C_{L_0}^2 \\
C_1 &= -2\bar{K}C_{L_0}C_{L_\alpha} \\
C_2 &= \bar{K}C_{L_\alpha}^2
\end{aligned}
\tag{8.81}
$$

Values of these quantities for the SBJ at $M = .6$ (App. A) are given by $C_0 = .0235$, $C_1 = .0629$, and $C_2 = 1.943$.

In Chap. 3, the induced drag of the horizontal tail was neglected relative to that of the wing in computing the drag polar. However, with a formula for the downwash angle at the horizontal tail, it is possible to calculate the induced drag of the horizontal tail and include it the drag polar.

8.12 Trimmed Drag Polar

Once the pitching moment is known, it is possible to include the effect of elevator deflection on the drag coefficient. Since the drag is essentially parabolic and the lift and aerodynamic pitching moment are linear, they can be expressed in the general forms

$$
\begin{aligned}
C_D &= \bar{C}_{D_0} + \bar{K}C_{l_\alpha}^2(\alpha - \alpha_{0L})^2 \tag{8.82} \\
C_L &= C_{L_0} + C_{L_\alpha}\alpha + C_{L_{\delta_E}}\delta_E \tag{8.83} \\
C_m^A &= C_{m_0}^A + C_{m_\alpha}^A\alpha + C_{m_{\delta_E}}^A\delta_E. \tag{8.84}
\end{aligned}
$$

At low subsonic speeds, the δ_E term in the lift can be neglected. Then, Eq. (8.83) can be solved for α in terms of C_L and used to eliminate α from Eq. (8.84). Since $C_{L_0} = -C_{L_\alpha}\alpha_{0L}$, the result has the form

$$C_D = \bar{C}_{D_0} + \bar{K}C_L^2, \tag{8.85}$$

as before.

If the δ_E terms are included, Eq. (8.84) is solved for the δ_E which makes $C_m^A = 0$ (trimmed flight), that is,

$$\delta_E = \frac{C_{m_0}^A}{C_{m_{\delta_E}}^A} - \frac{C_{m_\alpha}^A}{C_{m_{\delta_E}}^A}\alpha. \tag{8.86}$$

Then, δ_E is eliminated from Eq. (8.83) to obtain

$$C_L = C_{L_0}' + C_{L_\alpha}'\alpha \tag{8.87}$$

where

$$C_{L_0}' = C_{L_0} - C_{L_{\delta_E}}\frac{C_{m_0}^A}{C_{m_{\delta_E}}^A}, \quad C_{L_\alpha}' = C_{L_\alpha} - C_{L_{\delta_E}}\frac{C_{m_\alpha}^A}{C_{m_{\delta_E}}^A}. \tag{8.88}$$

Finally, α is eliminated from Eq. (8.82) which becomes

$$C_D = \bar{C}_{D_0} + \bar{K}\left(\frac{C_{L_\alpha}}{C_{L_\alpha}'}\right)^2 (C_L - C_{L_0}' - C_{L_\alpha}'\alpha_{0L})^2. \tag{8.89}$$

The *trimmed drag polar* has the form

$$C_D = C_{D_m} + K_m(C_L - C_{L_m})^2. \tag{8.90}$$

While this polar may be more accurate, its coefficients are now functions of cg position ($\bar{X}_{cg}$).

Problems

8.1 For an arbitrary planform (arbitrary $c(y)$), the equations for the mean aerodynamic chord and the location of the aerodynamic center are given by

$$\bar{c} = \frac{2}{S}\int_0^{\frac{b}{2}} c^2(y)\,dy, \quad \eta = \frac{2}{S}\int_0^{\frac{b}{2}} yc(y)\,dy, \quad \xi = \frac{2}{S}\int_0^{\frac{b}{2}} x(y)c(y)\,dy.$$

a. For a straight tapered wing (Figs. 3.8 and 8.5), show that

$$c(y) = c_r - (c_r - c_t)\frac{y}{b/2}, \quad x(y) = pc(y) + \tan \lambda_{le}\, y$$

where p is the airfoil ac/c.

b. Derive Eqs. (8.15) for a straight tapered wing.

8.2 Calculate the location of the aerodynamic center ξ, η and the mean aerodynamic chord $\bar{c}$ of the wing of the SBJ (App. A). Find the distance from the nose to the leading edge of the wing mean aerodynamic chord.

8.3 Calculate $\bar{X}_{ac_{WB}}$ and $\bar{X}_{ac_H}$ for the SBJ. Assuming that the cg is located 21.4 ft from the nose along x_b, show that $\bar{X}_{cg} = .30$.

8.4 Given all of the geometric data and the airfoil aerodynamic data listed in App. A and the results of Probs. 8.2 and 8.3, perform the tasks listed below for the SBJ operating at $M = 0.6$ and $\bar{X}_{cg}$:

a. Calculate $\alpha_{0LW}, C_{L\alpha_W}, C_{mac_W}, C_{L\alpha_H}$ and τ_E.

b. Calculate ε_α and ε_0.

c. Calculate $C_{L0}, C_{L\alpha}$, and $C_{L\delta_E}$.

d. Calculate $\bar{V}_H, C_{m0}^A, C_{m\alpha}^A, C_{m\delta_E}^A$, and $\bar{X}_{ac}$.

e. Calculate $C_{L_Q}, C_{m_Q}^A, C_{L\dot{\alpha}}, C_{m\dot{\alpha}}^A$

Once you have calculated these aerodynamic quantities and, hence, verified the results listed in App. A, use the results of App. A.

8.5 For the SBJ at $M = 0.6$ and $\bar{X}_{cg}$, show that the trimmed drag polar is given by

$$C_D = .023 + .0870(C_L + .00865)^2.$$

How does the trimmed drag polar compare with the untrimmed drag polar at this flight condition?

Chapter 9

Static Stability and Control

A result of trajectory analysis is the flight envelope of an airplane. The flight envelope is the region of the velocity-altitude plane where the airplane can operate in quasi-steady flight (climb, cruise, descent). However, in trajectory analysis only the forces acting on the airplane and the resulting motion of the center of gravity are considered. The next step is to analyze the forces and moments acting on the airplane and the combined motions of the center of gravity and about the center of gravity. It is important to determine whether or not a nominal pilot can control the airplane at all points of the flight envelope and for all center of gravity locations. Other issues concern the behavior of the airplane when subjected to control inputs and to wind gusts. This is the subject matter of stability and control.

With regard to stability, assume that an airplane is operating in cruise at a given flight condition. Then, at some time instant, the airplane is disturbed from this flight condition by a control surface deflection or a wind gust. The airplane is said to be dynamically stable if over time it returns to the original flight condition or to goes to a neighboring flight condition. On the other hand, static stability is concerned with what happens at the point of the disturbance. The airplane is said to be statically stable if it automatically produces forces and moments which tend to reduce the disturbance, that is, if certain quantities have the right signs. As an analogy, static stability is like point performance in trajectory analysis, and dynamic stability is like path performance.

With regard to control, there are two issues. In static control, the control force required to operate at a given flight condition is determined to ensure that it is within the capability of a nominal pilot. Also, the stick force should be zero in the neighborhood of the desired cruise speed, and it should take a push on the stick to increase the speed and a pull to decrease it. In dynamic control, the time-dependent response to a control input is examined.

It should be noted that an airplane can be unstable and still be flown safely providing the controls are sufficiently effective. This was the case with the Wright brothers airplane.

An important consideration is how the mass properties of the airplane (mass, cg, moment of inertia) affect its stability and control characteristics. Of particular concern is the effect of changing the cg position.

In addition to stability and control characteristics in quasi-steady flight. it is important to examine the behavior of an airplane in nonsteady flight. Particular trajectories include a pull-up in the vertical plane and a turn in the horizontal plane.

In general, small perturbation motion of an airplane causes the 6DOF equations of motion to divide into two sets. One set governs the pitching or *longitudinal motion* of the airplane. The other set governs the combined rolling and yawing motion or *lateral-directional motion* of the airplane. In this chapter, static longitudinal stability and control is discussed in some detail; only a few brief remarks are made about static lateral-directional stability and control.

9.1   Longitudinal Stability and Control

Several topics associated with static longitudinal stability and control for quasi-steady cruise, climb, and descent are discussed. First, equations for determining the angle of attack and the elevator angle (trim conditions) required to maintain a given flight condition are derived. Then, studying the effect of center of gravity position on trim leads to a forward cg limit. Next, speed stability and angle of attack stability are investigated. Angle of attack stability leads to an aft cg limit which is called the neutral point. The use of the elevator to control the airplane in pitch is discussed. Control characteristics of an airplane are studied with a view toward good handling qualities, that is, stick force and stick force gradient. Trim tabs are introduced to reduce the stick force to zero at a given flight condition. However, if the pilot trims the airplane and then lets go of the stick (free elevator), its stability characteristics are reduced. A trajectory which imposes additional control-related limits on cg position is the pull-up, leading to the maneuver point. Finally, a few brief remarks are made about static lateral-directional stability and control.

9.2 Trim Conditions for Steady Flight

The purpose of this section is to determine the angle of attack and elevator angle required for a given flight condition in a steady cruise, climb, or descent. For steady flight, it is assumed that $\dot{V}$, $\dot{\gamma}$, $\dot{Q}$, and $\dot{\alpha}$ are zero, that γ and $\varepsilon_0 + \alpha$ are small, and that $T(\varepsilon_0 + \alpha) << W$. Then, the dynamic equations for a steady cruise, climb, or descent are obtained from Eq. (8.4) as

$$
\begin{aligned}
T - D - W\gamma &= 0 \\
L - W &= 0 \\
M &= 0.
\end{aligned} \tag{9.1}
$$

These equations can be written in a nondimensional form by introducing force and moment coefficients. For a force F with a moment arm d, the force and moment coefficients are defined as

$$
C_F = \frac{F}{\bar{q}S}, \quad C_m^F = \frac{Fd}{\bar{q}S\bar{c}} \tag{9.2}
$$

where $\bar{q}$ is the dynamic pressure, S is the wing planform area, and $\bar{c}$ is the mean aerodynamic chord of the wing. The equations of motion become

$$
\begin{aligned}
C_T - C_D - C_W\gamma &= 0 \\
C_L - C_W &= 0 \\
C_m &= 0.
\end{aligned} \tag{9.3}
$$

Finally, if Eqs. (9.3), (8.44), and (8.61) are combined, the equations of motion become

$$
C_T - C_{D_0} - KC_L^2 - C_W\gamma = 0 \tag{9.4}
$$
$$
C_{L_0} + C_{L_\alpha}\alpha + C_{L_{\delta_E}}\delta_E - C_W = 0 \tag{9.5}
$$
$$
C_{m_0} + C_{m_\alpha}\alpha + C_{m_{\delta_E}}\delta_E = 0. \tag{9.6}
$$

If h, M, W, T are given, the flight path inclination can be determined from Eq. (9.4). On the other hand, if the flight path angle is given (as in level flight where $\gamma = 0$), Eq. (9.4) yields the thrust required to fly at that γ. This value of the thrust is used to determine the thrust moment. In either case, Eqs. (9.5) and (9.6) can be solved for α and δ_E as

$$\alpha = \frac{(C_L - C_{L_0})C_{m_{\delta_E}} + C_{m_0}C_{L_{\delta_E}}}{C_{L_\alpha}C_{m_{\delta_E}} - C_{m_\alpha}C_{L_{\delta_E}}} \tag{9.7}$$

and

$$\delta_E = -\frac{C_{L_\alpha}C_{m_0} + C_{m_\alpha}(C_L - C_{L_0})}{C_{L_\alpha}C_{m_{\delta_E}} - C_{m_\alpha}C_{L_{\delta_E}}} \tag{9.8}$$

where C_L is written in place of C_W, the two being equal. Because $C_m = 0$, these equations are referred to as *trim conditions* and can be used to ensure that the elevator is big enough to trim the airplane throughout the flight envelope. The trim conditions have been computed for the SBJ at the flight conditions of Sec. A.2 and are found to be α=2.23 deg and $\delta_E = 1.95$ deg.

For airplanes with all-moving horizontal tails (no elevator), trim is achieved by i_H. In this case Eqs. (9.5) and (9.6) can be solved for α and i_H.

Effect of cg position on trim conditions

The next step is to determine the effect of cg position on the trim conditions. The cg position is contained in all the C_m terms. However, since $\bar{V}_H$ does not change much with cg position, the major effect of cg position is contained in the C_{m_α} term as defined by Eq. (8.56). Also, in the denominator of Eqs. (9.7) and (9.8), the dominant term is $C_{L_\alpha}C_{m_{\delta_E}}$. As a consequence, cg position does not affect the trim angle of attack, but it does affect the trim elevator angle. Furthermore, as the cg is moved forward, the elevator angle becomes more negative. Since there is a limit on the magnitude of δ_E, there is a forward limit on cg position. The forward cg location is obtained from Eq. (9.8) by setting $\delta_E = \delta_{E_{min}}$ and is given by

$$\bar{X}_{cg} = \bar{X}_{ac} - \frac{C_{m_0} + C_{m_{\delta_E}}\delta_{E_{min}}}{C_L - C_{L_0}}. \tag{9.9}$$

Since it is desirable to control the airplane to $C_{L_{max}}$ at any Mach number, the least forward cg position occurs when $C_L = C_{L_{max}}$.

In general, the most restrictive forward cg limit occurs on landing approach with flaps down (highest $C_{L_{max}}$) and in ground effect where
the downwash angle at the horizontal tail is about half that in free flight.

9.3 Static Stability

An airplane is said to be statically stable if, following a disturbance,
forces and moments are produced by the airplane which *tend* to reduce
the disturbance. The word *tend* means that certain quantities have the
right sign.

Imagine an airplane in steady level flight, and assume that it
flies into a region of vertically ascending air whose effect is to instantaneously increase the speed of the airplane relative to the air and increase
the angle of attack. These effects are shown in Fig. 9.1. For the airplane
to be statically stable, it must generate forces and a moment which tend
to reduce the velocity disturbance and the angle of attack disturbance.

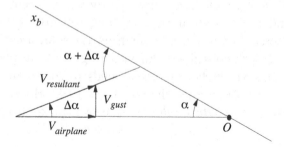

Figure 9.1: Increase in Velocity and Angle of Attack Due to Gust

In level flight (cruise), the principle effect of a speed disturbance is governed by the equation

$$m\dot{V} = T - D. \tag{9.10}$$

where at the trim point

$$\dot{V} = 0, \quad T - D = 0. \tag{9.11}$$

If the disturbance is such that the speed is increased by a small amount,
$T - D$ must become negative so that $\dot{V} < 0$ and the velocity disturbance

tends to decrease. If the speed is decreased, $T - D$ must become positive so that $\dot{V} > 0$ and the velocity disturbance tends to increase. Note that these conditions are satisfied at the high-speed solution (see Fig. 4.1) but not at the low-speed solution. Hence, the high-speed solution is speed stable, but the low-speed solution is not.

The principal effect of an angle of attack disturbance is governed by the equation

$$I_{yy}\dot{Q} = M \tag{9.12}$$

where at the trim point

$$\dot{Q} = 0, \quad M = 0. \tag{9.13}$$

If the disturbance is such that the angle of attack increases, M must become negative so that $\dot{Q} < 0$ and the angle of attack disturbance tends to decrease. If the angle of attack is decreased, M must become positive so that $\dot{Q} > 0$.

In coefficient form, Fig. 9.2 shows a plot of pitching moment coefficient versus angle of attack in which the moment curve slope C_{m_α} is negative and C_m is zero at the trim angle of attack. If the angle of attack is increased a small amount from the trim angle of attack a small amount $\Delta\alpha$, the pitching moment acting on the airplane becomes negative. A negative pitching moment tends to rotate the nose of the airplane downward and reduce the angle of attack. Hence, an airplane with $C_{m_\alpha} < 0$ is *statically stable* in angle of attack. If $C_{m_\alpha} > 0$, an

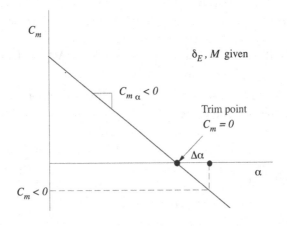

Figure 9.2: Static Stability

increase in α causes a positive pitching moment (nose up) which tends to increase α still further. Such an airplane is *statically unstable*. Further, if $C_{m_\alpha} = 0$, an increase in α does not change C_m (it is still zero), and the angle of attack stays at the perturbed value. This airplane is *statically neutrally stable*.

An airplane without a horizontal tail is statically unstable, and it is the addition of the horizontal tail that makes it stable. The amount of static stability wanted for the airplane, that is, the value of C_{m_α}, determines the location and size of the horizontal tail.

Effect of cg position on angle of attack stability

From Eqs. (8.58) and (8.62), it is known that

$$C_{m_\alpha} = C_{L_\alpha}(\bar{X}_{cg} - \bar{X}_{ac}) \qquad (9.14)$$

where $C_{L_\alpha} > 0$. Hence, if $\bar{X}_{cg} < \bar{X}_{ac}$, the airplane is statically stable $(C_{m_\alpha} < 0)$. If $\bar{X}_{cg} = \bar{X}_{ac}$, the airplane is statically neutrally stable $(C_{m_\alpha} = 0)$, Finally, if $\bar{X}_{cg} > \bar{X}_{ac}$, the airplane is statically unstable. For an airplane to be inherently (by design) aerodynamically statically stable, the center of gravity must be ahead of the aerodynamic center. Since the aerodynamic center is the cg location where the airplane is neutrally stable, it is also called the *neutral point*. The distance between the center of gravity and the aerodynamic center or neutral point, that is, $\bar{X}_{ac} - \bar{X}_{cg}$, is called the *static margin*. Static margin is an important design parameter for inherently aerodynamically stable airplanes.

The Wright brother's airplane was aerodynamically unstable, but they were able to fly their airplane because of its control system. Once the application of aerodynamic stability was understood, airplanes were designed to be aerodynamically stable. This design criterion (positive static margin) is still used today with the exception of modern jet fighters beginning with the F-16. Designed to be a lightweight fighter, the F-16 was allowed to be aerodynamically statically unstable (subsonically) to decrease the size of its horizontal tail and, hence, its weight. The F-16 is made stable, however, by its automatic flight control system. A rate gyro senses an uncommanded pitch rate (a pitch rate caused by a gust, for example), and the control system deflects the elevator almost instantaneously to generate an opposing pitching moment to cancel out the pitch rate.

9.4   Control Force and Handling Qualities

The air flowing over an elevator creates a pressure distribution on its surface that causes a moment about the elevator hinge line called the elevator hinge moment. To keep the elevator at a particular angle, a control moment opposite in sign to the hinge moment must be provided by the pilot in form of a force on the control column.

In the process of designing an airplane, the force required of the pilot must be kept within his physical capability. The magnitude of this pilot force depends on the size of the airplane and the speed for which it is designed. For small, slow airplanes, the pilot is connected directly to the elevator (reversible control system), and he feels the full effect of the moment acting on the elevator. For large, slow airplanes such as the B-52, the elevator may be so heavy that the pilot cannot move it by himself. In this case the airplane is designed with a hydraulic or electrical system which provides part or even all the necessary control moment. This is also true for small and large, fast airplanes where the aerodynamic hinge moment becomes very large due to high dynamic pressure.

For the case where the hydraulic or electrical system provides all of the control moment (irreversible control system), the pilot commands an elevator deflection, and the automatic flight control system causes the elevator to be deflected to that angle. To give the pilot some feel about how much force is being created, an artificial feel system is provided. This artificial feel can be created by a spring of controlled stiffness opposing the rotation of the control column.

To illustrate the analysis of stick force, hinge moment, etc., consider the reversible control system (no power assist) shown in Fig. 9.3 where the region marked gearing represents the mechanical linkage of the control system, and assume that the airplane is flying at a low subsonic speed (Mach number effects negligible in the aerodynamics). A positive stick force is a pull, and a positive stick deflection is aft. At the elevator, a positive hinge moment is clockwise, same as the pitching moment. The *stick force* can be expressed as

$$F_s = GH \qquad (9.15)$$

where $G > 0$ is the gearing and H is the aerodynamic hinge moment. The gearing includes the moment arm which converts the stick force to a moment.

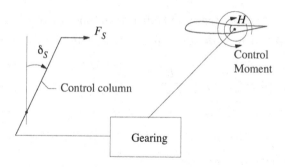

Figure 9.3: Sign Conventions for Control

The *elevator hinge moment* is converted to coefficient form by dividing by the local dynamic pressure $\bar{q}_H$, the elevator area S_E aft of the hinge line, and an average chord c_E of the elevator aft of the chord line, that is,

$$C_h = \frac{H}{\bar{q}_H S_E \bar{c}_E} \tag{9.16}$$

so the stick force becomes

$$F_s = G\bar{q}_H S_E \bar{c}_E C_h. \tag{9.17}$$

Next, the hinge moment coefficient is assumed to vary linearly with α_H and δ_E so that

$$C_h = C_{h_0} + C_{h_{\alpha_H}}\alpha_H + C_{h_{\delta_E}}\delta_E \tag{9.18}$$

where $C_{h_0} = 0$ because tail surfaces are symmetric. From Eqs. (8.39), (8.40), and (9.17), it is seen that

$$F_s = G\bar{q}\eta_H S_E \bar{c}_E \left[C_{h_{\alpha_H}}(i_H - \varepsilon_0) + C_{h_{\alpha_H}}(1 - \varepsilon_\alpha)\alpha + C_{h_{\delta_E}}\delta_E \right]. \tag{9.19}$$

Then, to isolate the dynamic pressure, Eq. (9.19) is combined with Eqs. (9.7), (9.8), and $C_L = W/\bar{q}S$, leading to the following result for the stick force:

$$F_S = A + BV^2 \tag{9.20}$$

where A and B are independent of V. The specific forms for A and B are easily derived if needed.

Eq. (9.20) shows how control force varies with velocity. In terms of velocity, the stick force should have the form shown in Fig. 9.4 for good airplane *handling qualities*. The stick force should be zero at a speed near the desired cruise speed; the maximum required stick force should be within the capability of a nominal pilot; and the *stick force gradient*

$$\frac{\partial F_s}{\partial V} = 2\, BV \qquad (9.21)$$

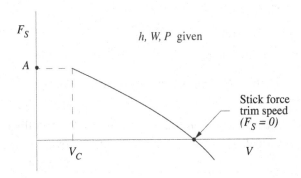

Figure 9.4: Stick Force versus Flight Speed

should be negative at this point. Hence, to increase the speed away from the stick force trim speed $(F_s = 0)$, a push stick force is required, and to decrease the speed, a pull is needed. These handling qualities are obtained if $A > 0$ and $B < 0$ and if the maximum stick force is within pilot capability $(A < A_{max})$. In Fig. 9.4, V_C is the minimum speed at with the elevator can rotate the airplane.

It is possible to achieve $F_s = 0$ at the desired flight speed with a trim tab. It is also possible to do this with a movable horizontal stabilizer through i_H, if this control is available.

9.5   Trim Tabs

At an arbitrary flight condition, the stick force is not zero, and to operate the airplane at the given speed, the pilot must produce a continuous stick force. Trim tabs were invented to cause the elevator to float $(C_h = 0)$ at a given angle so that the stick force can be made zero over a range of flight speeds. A *trim tab*(see Fig. 9.5) is a small flap on the trailing edge of a control surface; see the ailerons of the subsonic business jet in App. A. When deflected, trim tabs have a negligible effect on lift and

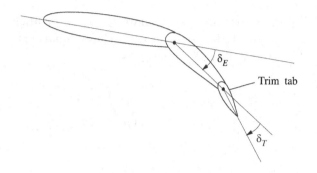

Figure 9.5: Trim Tab

drag but produce a moment about the hinge line which can offset the elevator hinge moment.

To analyze the effect of a trim tab, the hinge moment is written as the linear relation

$$C_h = C_{h_{\alpha_H}} \alpha_H + C_{h_{\delta_E}} \delta_E + C_{h_{\delta_T}} \delta_T \tag{9.22}$$

where $C_{h_0} = 0$ for symmetric tails. If the trim tab is used to make $C_h = 0$ (zero stick force), the elevator floats at the angle

$$\delta_E = -\frac{C_{h_{\alpha_H}} \alpha_H + C_{h_{\delta_T}} \delta_T}{C_{h_{\delta_E}}} \tag{9.23}$$

Combined with Eqs. (8.39) and (8.40), the elevator angle is related to the angle of attack and the trim tab angle as

$$\delta_E = -\frac{C_{h_{\alpha_H}}}{C_{h_{\delta_E}}} [i_H - \varepsilon_0 + (1 - \varepsilon_\alpha)\alpha] - \frac{C_{h_{\delta_T}}}{C_{h_{\alpha_H}}} \delta_T. \tag{9.24}$$

Note that the angle at which the elevator floats depends on the angle of attack. If α changes, δ_E will change for the given trim tab setting.

Consider the situation in which the pilot trims the airplane at a given speed to obtain zero stick force. If the pilot holds the control so that the elevator cannot move, the stability characteristics do not change. On the other hand, if the pilot lets go of the control, the elevator is free to move if the angle of attack changes. In addition, the stability characteristics of the airplane change.

If Eq. (9.24) is used to eliminate the elevator angle from Eqs. (8.44) and (8.53), all of the coefficients of these equations change. In particular, C_{m_α} (now called C'_{m_α}) becomes

$$
\begin{aligned}
C'_{m_\alpha} &= C_{L_{\alpha WB}}(\bar{X}_{cg} - \bar{X}_{ac_{WB}}) \\
&- C_{L_{\alpha H}}(1 - \varepsilon_\alpha)\eta_H \bar{V}_H \left(1 - \tau_E \frac{C_{h_{\alpha H}}}{C_{h_{\delta E}}}\right).
\end{aligned}
\tag{9.25}
$$

The term

$$
1 - \tau_E \frac{C_{h_{\alpha H}}}{C_{h_{\delta E}}}
\tag{9.26}
$$

reduces the horizontal tail effectiveness. In addition, the aerodynamic center or neutral point of the airplane ($C'_{m_\alpha} = 0$) becomes

$$
\bar{X}'_{ac} = \frac{C_{L_{\alpha WB}}\bar{X}_{ac_{WB}} + C_{L_{\alpha H}}(1 - \varepsilon_\alpha)\eta_H \frac{S_H}{S}\left(1 - \tau_E \frac{C_{h_{\alpha H}}}{C_{h_{\delta E}}}\right)\bar{X}_{ac_H}}{C_{L_{\alpha WB}} + C_{L_{\alpha H}}(1 - \varepsilon_\alpha)\eta_H \frac{S_H}{S}\left(1 - \tau_E \frac{C_{h_{\alpha H}}}{C_{h_{\delta E}}}\right)}
\tag{9.27}
$$

This neutral point is called the *stick-free neutral point* and the previous neutral point (8.57) is often referred to as the *stick-fixed neutral point*. In general, allowing the elevator to float freely causes the neutral point to move forward, reducing the stability of the airplane.

9.6   Trim Conditions for a Pull-up

The reference path used in the previous derivations is a quasi-steady climb, cruise or descent. Another reference path that is used to investigate control is the a pull-up, which is an accelerated maneuver. An important aerodynamic control characteristic is the elevator displacement required to make an n-g pull-up or the *elevator angle per g*. There exists a cg position where the elevator angle per g goes to zero, making it too easy to make an high-g maneuver and destroy the airplane. This cg position is behind the neutral point and is called the maneuver point, a formula for which is derived here.

Consider a constant power setting, constant angle of attack pull-up. The 6DOF equations of motion in wind axes are listed in Eqs. (8.4) and (8.5). If it is assumed that $\dot{V}, \dot{Q}, \varepsilon$, and $T\varepsilon$ are small and that the pull-up is to be investigated at its low point where $\gamma = 0$, the dynamic equations become

$$
\begin{aligned}
0 &= T - D \\
Q &= (g/WV)(L - W) \\
0 &= M.
\end{aligned}
\tag{9.28}
$$

In terms of nondimensional coefficients (see Sec. 9.2), these equations can be rewritten as

$$
\begin{aligned}
0 &= C_T - C_{D_0} - KC_W^2 n^2 \\
\hat{Q} &= (\bar{c}g/2V^2)(n - 1) \\
n &= C_L/C_W \\
0 &= C_m.
\end{aligned}
\tag{9.29}
$$

The quantity

$$
\hat{Q} = \bar{c}Q/2V
\tag{9.30}
$$

is the nondimensional pitch rate. Since $\dot{\alpha} = 0$, it is known from Sec. 8.2 that

$$
\begin{aligned}
C_L &= C_{L_0} + C_{L_\alpha}\alpha + C_{L_{\delta_E}}\delta_E + C_{L_Q}\hat{Q} \\
C_m &= C_{m_0} + C_{m_\alpha}\alpha + C_{m_{\delta_E}}\delta_E + C_{m_Q}\hat{Q}
\end{aligned}
\tag{9.31}
$$

where

$$
C_{m_0} = C_{m_0}^T + C_{m_0}^A, \quad C_{m_\alpha} = C_{m_\alpha}^A, \quad C_{m_{\delta_E}} = C_{m_{\delta_E}}^A, \quad C_{m_Q} = C_{m_Q}^A.
\tag{9.32}
$$

Formulas for calculating the Q derivatives are given in Sec. 8.9.

Given h, M, W, P and n, the first of Eqs. (9.29) gives the thrust which is used to compute the thrust moment. The remaining three equations lead to

$$
\begin{aligned}
C_{L_0} + C_{L_\alpha}\alpha + C_{L_{\delta_E}}\delta_E + C_{L_Q}\hat{Q} &= C_W n \\
C_{m_0} + C_{m_\alpha}\alpha + C_{m_{\delta_E}}\delta_E + C_{m_Q}\hat{Q} &= 0
\end{aligned}
\tag{9.33}
$$

where

$$
\hat{Q} = (g\bar{c}/2V^2)(n - 1).
\tag{9.34}
$$

These equations are solved for α and δ_E as follows:

$$
\alpha = \frac{C_{m_{\delta_E}}(C_W n - C_{L_0} - C_{L_Q}\hat{Q}) + C_{L_{\delta_E}}(C_{m_0} + C_{m_Q}\hat{Q})}{C_{L_\alpha}C_{m_{\delta_E}} - C_{m_\alpha}C_{L_{\delta_E}}}
\tag{9.35}
$$

$$\delta_E = -\frac{C_{L_\alpha}(C_{m_0} + C_{m_Q}\widehat{Q}) + C_{m_\alpha}(C_W n - C_{L_0} - C_{L_Q}\widehat{Q})}{C_{L_\alpha}C_{m_{\delta_E}} - C_{m_\alpha}C_{L_{\delta_E}}}. \qquad (9.36)$$

There exists a cg position where the elevator angle becomes independent of n, meaning that a small elevator deflection can create a large n maneuver. This cg position comes from the elevator angle per g, that is,

$$\frac{\partial \delta_E}{\partial n} = -\frac{C_{L_\alpha}C_{m_Q}(\bar{c}g/2V^2) + C_{m_\alpha}[C_W - C_{L_Q}(\bar{c}g/2V^2)]}{C_{L_\alpha}C_{m_{\delta_E}} - C_{m_\alpha}C_{L_{\delta_E}}}. \qquad (9.37)$$

Using Eq. (8.58) for C_{m_α} and setting the elevator angle per g to zero leads to the following equation for the *maneuver point*:

$$\bar{X}_{cg} = \bar{X}_{ac} - \frac{C_{m_Q}(\bar{c}g/2V^2)}{C_W - C_{L_Q}(\bar{c}g/2V^2)}. \qquad (9.38)$$

Numerically, the second term in the denominator is negligible with respect to the first. Then, because $C_{m_Q} < 0$, the maneuver point is aft of the neutral point. While an airplane can have the cg behind the ac with stability provided electronically, the cg must be ahead of the maneuver point.

There is another maneuver point associated with the stick force per g going to zero at some cg position (see Ref. ER). It is called the *stick-free maneuver point*, and it is aft of the *stick-fixed maneuver point* (9.38).

It is important to know where the neutral points and the maneuver points are so that the aft bound on the cg position can be chosen a safe distance from all of them.

9.7 Lateral-Directional Stability and Control

In summary, static longitudinal stability and control involves the force coefficient in the x direction C_x, the force coefficient in the z direction C_z, and the pitching moment coefficient C_m. Each of these items is a function of angle of attack α, Mach number M, and elevator deflection δ_E. The control is the elevator, and pitch stability is provided by the horizontal tail, requiring that $C_{m_\alpha} < 0$. Recall that C_{m_α} is the slope of the C_m versus α line.

Static lateral-directional stability and control involves the side force coefficient C_y, the yawing moment coefficient C_n, and the rolling moment coefficient C_l. These quantities are functions of the sideslip angle β (see Fig. 9.6), the Mach number M, the aileron deflection δ_A, and the rudder deflection δ_R. Rolling motion is controlled essentially by the

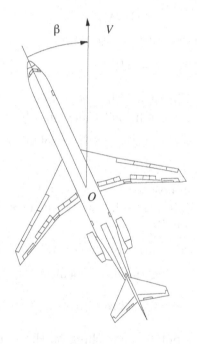

Figure 9.6: Sideslip Angle

ailerons, and roll or lateral stability is provided by wing dihedral, wing location on the fuselage (high or low), and wing sweep. It requires that $C_{l_\beta} < 0$, where C_{l_β} is the slope of the C_l versus β line. Yawing motion is controlled essentially by the rudder. Yaw or directional stability is provided by the vertical tail and requires that $C_{n_\beta} > 0$. C_{n_β} is the slope of the C_n versus β line.

For multi-engine airplanes, one-engine-out trimmed flight (no moments) requires both aileron and rudder deflections. Furthermore, this trim condition sizes the vertical stabilizer.

Problems

9.1 At this point, all of the geometric data and the aerodynamic data
are known for the SBJ (App. A). Assume that the SBJ is operating
at the following cruise conditions:

$$h = 30,000 \text{ ft}, \quad \rho = 0.000889 \text{ slug/ft}^2, \quad a = 995 \text{ ft/s}^2$$

$$V = 597 \text{ ft/s}, \quad W = 11,000 \text{ lb}, \quad \gamma = 0.0 \text{ deg}$$

$$M = 0.6, \quad C_L = .299, \quad \bar{X}_{cg} = .300$$

$$i_H = -2.0 \text{ deg}, \quad l_T = -2.0 \text{ ft}$$

$$\bar{C}_{D_0} = 0.023, \quad \bar{K} = 0.073.$$

a. How much thrust are the engines producing in this constant
altitude flight condition? Calculate C_T and $C_{m_0}^T$.

b. At what angle of attack is the airplane operating?

c. What is the elevator angle?

d. What is the most forward cg position for this flight condition
if $\delta_{E_{\min}} = -20$ deg. What would the most forward cg position
be if C_L were $C_{L_{\max}} = 1.2$?

e. Is the aircraft statically stable? What is the static margin?

f. What is the hinge moment coefficient if

$$C_{h_{\alpha_H}} = -0.383 \text{ rad}^{-1}, \quad C_{h_{\delta_E}} = -0.899 \text{ rad}^{-1}.$$

Is the pilot pulling or pushing on the control column?

g. If the SBJ had elevator trim tabs and $C_{h_{i_H}} = -.776$, what trim
tab angle is needed to get the elevator to float at the angle of
part c. Does the sign of the trim tab deflection make sense?

h. Calculate the location of the stick-fixed maneuver point.

9.2 For an airplane with an all-moving horizontal tail (no elevator),
derive the equations for the angle of attack and the incidence of
the horizontal tail for trim, that is,

$$\alpha = \frac{(C_L - \bar{C}_{L_0})C_{m_{i_H}} + \bar{C}_{m_0}C_{L_{i_H}}}{C_{L_\alpha}C_{m_{i_H}} - C_{m_\alpha}C_{L_{i_H}}}$$

$$i_H = -\frac{C_{L_\alpha}\bar{C}_{m_0} + C_{m_\alpha}(C_L - \bar{C}_{L_0})}{C_{L_\alpha}C_{m_{i_H}} - C_{m_\alpha}C_{L_{i_H}}}$$

where $\bar{C}_{L_0}$ is that part of C_{L_0} that does not contain i_H, $\bar{C}_{m_0}$ is that part of C_{m_0} that does not contain i_H, and

$$C_{L_{i_H}} = C_{L_{\alpha_H}}\eta_H\frac{S_H}{S}, \quad C_{m_{i_H}} = -C_{L_{\alpha_H}}\eta_H\bar{V}_H.$$

Assume that the SBJ has an all-moving horizontal tail. For the flight conditions of Prob. 9.1, calculate α and i_H.

Chapter 10

6DOF Model: Body Axes

While the analysis of dynamic stability and control can be carried out in wind axes, the convention is to use the equations of motion in the body axes. After the equations of motion are derived in the regular body axes system, they are derived in the stability axes system. The stability axes are a set of body axes that are attached to the airplane at an angle relative to the regular body axes. Stability axes are used in the study of dynamic stability and control.

10.1 Equations of Motion: Body Axes

The assumptions and the coordinate systems are the same as those of Sec. 2.1 with the exception of discarding the moments when forces are moved to the center of gravity. Fig. 10.1 shows the *regular body axes* $O_{x_b y_b z_b}$ which are fixed to the aircraft. The body axes are orientated relative to the local horizon by the pitch angle Θ . Hence, the unit vectors of these two systems satisfy the relations

$$
\begin{aligned}
\mathbf{i}_b &= \cos\Theta \mathbf{i}_h - \sin\Theta \mathbf{k}_h \\
\mathbf{k}_b &= \sin\Theta \mathbf{i}_h + \cos\Theta \mathbf{k}_h.
\end{aligned}
\tag{10.1}
$$

Then, since the local horizon unit vectors are constant, the time derivatives of the body axes unit vectors are given by

$$
\begin{aligned}
\frac{d\mathbf{i}_b}{dt} &= -\dot{\Theta}\mathbf{k}_b \\
\frac{d\mathbf{k}_b}{dt} &= \dot{\Theta}\mathbf{i}_b.
\end{aligned}
\tag{10.2}
$$

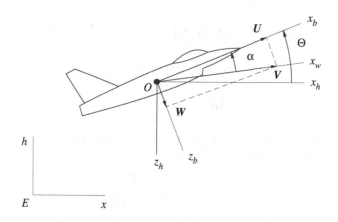

Figure 10.1: Body Axes System

Fig. 10.1 also shows the velocity vector projected onto the body axes. Because W is now used to denote a component of the velocity vector, the weight is expressed as mg. The velocity vector can be written as

$$\mathbf{V} = U\mathbf{i}_b + W\mathbf{k}_b. \tag{10.3}$$

As a result the magnitude of the velocity vector is given by

$$V = \sqrt{U^2 + W^2}, \tag{10.4}$$

and the angle of attack satisfies the relation

$$\tan \alpha = \frac{W}{U}. \tag{10.5}$$

10.1.1 Translational kinematic equations

The translational kinematic equations result from the definition of the velocity:

$$\mathbf{V} = \frac{d\mathbf{EO}}{dt}. \tag{10.6}$$

The scalar equations are obtained by writing the vectors in the local horizon system where the unit vectors are constant. Combining Eqs.

(10.1) and (10.3) leads to

$$\mathbf{V} = (U\cos\Theta + W\sin\Theta)\mathbf{i}_h + (-U\sin\Theta + W\cos\Theta)\mathbf{k}_h. \qquad (10.7)$$

Next, since the position vector is given by

$$\mathbf{EO} = x\mathbf{i}_h - h\mathbf{k}_h, \qquad (10.8)$$

the definition of velocity leads to the scalar equations

$$\begin{aligned}
\dot{x} &= U\cos\Theta + W\sin\Theta \\
\dot{h} &= U\sin\Theta - W\cos\Theta.
\end{aligned} \qquad (10.9)$$

10.1.2 Translational dynamic equations

The translational dynamic equations are derived from Newton's second law $\mathbf{F} = m\mathbf{a}$ and the definition of inertial acceleration $\mathbf{a} = d\mathbf{V}/dt$. From the expression (10.3) for the velocity, the acceleration is given by

$$\mathbf{a} = \dot{U}\mathbf{i}_b + U\frac{d\mathbf{i}_b}{dt} + \dot{W}\mathbf{k}_b + W\frac{d\mathbf{k}_b}{dt} \qquad (10.10)$$

which with the unit vector rates (10.2) becomes

$$\mathbf{a} = (\dot{U} + W\dot{\Theta})\mathbf{i}_b + (\dot{W} - U\dot{\Theta})\mathbf{k}_b. \qquad (10.11)$$

Next, from Fig. 10.2, it is seen that

$$\begin{aligned}
\mathbf{F} &= [\,T\cos\varepsilon_0 + L\sin\alpha - D\cos\alpha - mg\sin\Theta\,]\mathbf{i}_b \\
&+ [-T\sin\varepsilon_0 - L\cos\alpha - D\sin\alpha + mg\cos\Theta\,]\mathbf{k}_b
\end{aligned} \qquad (10.12)$$

Combining Eqs. (10.1), (10.12) with $\mathbf{F} = m\mathbf{a}$ leads to the translational dynamic equations

$$\begin{aligned}
m(\dot{U} + W\dot{\Theta}) &= T\cos\varepsilon_0 + L\sin\alpha - D\cos\alpha - mg\sin\Theta \\
m(\dot{W} - U\dot{\Theta}) &= -T\sin\varepsilon_0 - L\cos\alpha - D\sin\alpha + mg\cos\Theta.
\end{aligned} \qquad (10.13)$$

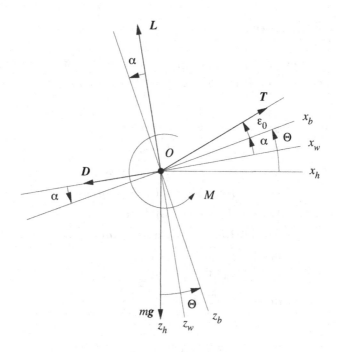

Figure 10.2: Forces and Moment

10.1.3 Rotational kinematic and dynamic equations

The equations for rotational motion of the airplane about the y_b axis are the same as those for rotational motion about the y_w axis. Hence, from Sec. 8.1, it is known that the rotational kinematic and dynamic equations are given by

$$
\begin{aligned}
\dot{\Theta} &= Q \\
I_{yy}\dot{Q} &= M^T + M^A.
\end{aligned}
\tag{10.14}
$$

10.1.4 Mass equations

As in Chap. 2, the mass change is governed by the equation

$$
\dot{m}g = -CT.
\tag{10.15}
$$

where the mass is used because W now denotes a velocity component. Also, the mass moment of inertia I_{yy} is assumed to be a known function of the mass.

10.1.5 Summary

The six degree of freedom equations of motion for nonsteady flight in a vertical plane over a flat earth in regular body axes are given by

$$
\begin{aligned}
\dot{x} &= U\cos\Theta + W\sin\Theta \\
\dot{h} &= U\sin\Theta - W\cos\Theta \\
\dot{U} &= -WQ + (1/m)[\,T\cos\varepsilon_0 \\
&\quad + \; L\sin\alpha - D\cos\alpha - mg\sin\Theta] \\
\dot{W} &= \quad UQ - (1/m)[T\sin\varepsilon_0 \\
&\quad + \; L\cos\alpha + D\sin\alpha - mg\cos\Theta] \\
\dot{\Theta} &= Q \\
\dot{Q} &= (M^A + M^T)/I_{yy} \\
\dot{m} &= -CT/g.
\end{aligned}
\tag{10.16}
$$

where

$$
V = \sqrt{U^2 + W^2}, \quad \tan\alpha = \frac{W}{U}.
\tag{10.17}
$$

These equations contain several quantities (D, L, T, C) which are functions of variables already present. The propulsion and aerodynamic terms in the equations of motion are assumed to satisfy the following functional relations:

$$
\begin{aligned}
T &= T(h, V, P), \quad C = C(h, V, P) \\
D &= D(h, V, \alpha), \quad L = L(h, V, \alpha, \delta_E, Q, \dot{\alpha}) \\
M^T &= M^T(h, V, P, \alpha), \quad M^A = M^A(h, V, \alpha, \delta_E, Q, \dot{\alpha}).
\end{aligned}
\tag{10.18}
$$

where δ_E is the elevator angle.

Eqs. (10.16) involve nine variables (seven states x, h, U, W, Θ, Q, m and two controls P and δ_E). Hence, there are two mathematical degrees of freedom. To solve these equations, three additional equations involving existing variables must be provided. An example is specifying the three control histories $P = P(t)$ and $\delta_E(t)$, which are the controls available to the pilot, depending on the design of the horizontal tail.

Formulas for computing the aerodynamics of Eq. (10.18) have already been derived in Chap. 8.

10.2 Equations of Motion: Stability Axes

The stability body axes system $O_{x_s y_s z_s}$ is shown in Fig. 10.3. *Stability body axes* are body axes system that are fixed to the airplane at a different orientation than the regular body axes. Note that U, W, α and Θ are

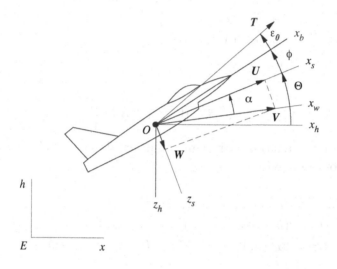

Figure 10.3: Stability Axes System

now defined relative to the stability axes. However, the aerodynamics is defined relative to the regular body axes, that is, the regular angle of attack. If ϕ denotes the angle between the regular body axes and the stability axes, the regular angle of attack, now denoted by $\bar{\alpha}$, is given by

$$\bar{\alpha} = \alpha + \phi. \tag{10.19}$$

The 6DOF equations of motion in stability axes for flight in a vertical plane can be derived in the same manner as Eqs. (10.16) and are given by

$$
\begin{aligned}
\dot{x} &= U\cos\Theta + W\sin\Theta \\
\dot{h} &= U\sin\Theta - W\cos\Theta \\
\dot{U} &= -WQ + (1/m)[\,T\cos(\varepsilon_0 + \phi) \\
        &\quad +\ L\sin\alpha - D\cos\alpha - mg\sin\Theta] \\
\dot{W} &= UQ - (1/m)[T\sin(\varepsilon_0 + \phi) \\
        &\quad +\ L\cos\alpha + D\sin\alpha - mg\cos\Theta] \\
\dot{\Theta} &= Q \\
\dot{Q} &= (M^A + M^T)/I_{yy} \\
\dot{m} &= -CT/g
\end{aligned}
\tag{10.20}
$$

where
$$
V = \sqrt{U^2 + W^2}, \quad \tan\alpha = \frac{W}{U}.
\tag{10.21}
$$

The functional relations for engine and aerodynamic behavior are obtained from Eq. (8.6) as

$$
\begin{aligned}
T &= T(h, V, P), \quad C = C(h, V, P) \\
D &= D(h, V, \bar{\alpha}), \quad L = L(h, V, \bar{\alpha}, \dot{\bar{\alpha}}, Q, \delta_E) \\
M^T &= M^T(h, V, P), \quad M^A = M^A(h, V, \bar{\alpha}, \dot{\bar{\alpha}}, Q, \delta_E).
\end{aligned}
\tag{10.22}
$$

Finally, the equations of motion in stability axes have two mathematical degrees of freedom, the same as those of the regular body axes.

10.3   Flight in a Moving Atmosphere

From the derivation of the equations of motion in the wind axes (Sec. 2.9), it is seen that the translational equations are given by

$$
\frac{d\mathbf{EO}}{dt} = \mathbf{V}_0, \quad \mathbf{F} = m\mathbf{a}_0, \quad \mathbf{a}_0 = \frac{d\mathbf{V}_0}{dt}.
\tag{10.23}
$$

In these equations, the inertial velocity $\mathbf{V}_0$ is written as the vector sum of the velocity relative to the atmosphere $\mathbf{V}$ plus the velocity of the atmosphere relative to the ground $\mathbf{w}$, that is,

$$
\mathbf{V}_0 = \mathbf{V} + \mathbf{w}
\tag{10.24}
$$

As a consequence, the translational equations of motion become

$$\frac{d\mathbf{EO}}{dt} = \mathbf{V} + \mathbf{w}, \quad \frac{d\mathbf{V}}{dt} = \frac{\mathbf{F}}{m} - \frac{d\mathbf{w}}{dt}. \tag{10.25}$$

As long as the reference point for moments is the center of gravity, the moving atmosphere has no effect on the rotational equations.

The derivation of the stability axes equations is the same as that in Sec. 10.2 with the added moving atmosphere terms. Hence, the equations of motion for flight in a moving atmosphere in stability axes are given by

$$
\begin{aligned}
\dot{x} &= U\cos\Theta + W\sin\Theta + w_x \\
\dot{h} &= U\sin\Theta - W\cos\Theta + w_h \\
\dot{U} &= -WQ + (1/m)[\,T\cos(\varepsilon_0 + \phi) \\
&\quad + L\sin\alpha - D\cos\alpha - mg\sin\Theta] \\
&\quad - \dot{w}_x\cos\Theta - \dot{w}_h\sin\Theta \\
\dot{W} &= UQ - (1/m)[T\sin(\varepsilon_0 + \phi) \\
&\quad + L\cos\alpha + D\sin\alpha - mg\cos\Theta] \\
&\quad - \dot{w}_x\sin\Theta + \dot{w}_h\cos\Theta \\
\dot{\Theta} &= Q \\
\dot{Q} &= (M^A + M^T)/I_{yy} \\
\dot{m} &= -CT/g
\end{aligned}
\tag{10.26}
$$

Problems

10.1 Consider the motion of a rocket in a vertical plane over a flat earth outside the atmosphere (see Fig. 10.4). The orientation of the rocket relative to the earth is given by the pitch angle $\theta(t)$. The orientation of the thrust relative to the rocket centerline is given by the gimbal angle $\delta(t)$ of the engine, and l_T is the distance from the center of gravity O to the base of the rocket. Derive the 6DOF equations of motion. Show that

$$
\begin{aligned}
\ddot{x} &= \frac{T}{m}\cos(\theta + \delta) \\
\ddot{h} &= \frac{T}{m}\sin(\theta + \delta) - g
\end{aligned}
$$

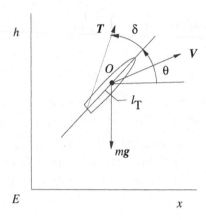

Figure 10.4: Rocket in Flight

$$\ddot{\theta} = \frac{T \sin \delta \, l_T}{I_{yy}}$$
$$\dot{m} = -\beta$$

where $T = \beta c$, c is the engine exhaust velocity assumed constant, β is the propellant mass flow rate, and I_{yy} is the mass moment of inertia about the pitch axis.

This problem provides another example of uncoupling the force and moment equations to form a 3DOF model. At this point, the gimbal angle $\delta(t)$ is the control. By assuming that the gimbal angle is small, the equations of motion can be rewritten as

$$\ddot{x} = \frac{T}{m} \cos \theta$$
$$\ddot{h} = \frac{T}{m} \sin \theta - g$$
$$\ddot{\theta} = \frac{T \delta \, l_T}{I_{yy}}$$
$$\dot{m} = -\beta$$

With these equations, the $\ddot{\theta}$ equation can be uncoupled from the system, and θ can be used as the control variable for the remaining equations. Once $\theta(t)$ has been determined, the $\ddot{\theta}$ equation can be used to compute the $\delta(t)$ needed to produce $\theta(t)$.

Chapter 11

Dynamic Stability and Control

The purpose of this chapter is to study the translational and rotational motion of an airplane to determine its dynamic stability and control characteristics. While it is possible to do this by numerically integrating the nonlinear six degree of freedom equations of motion, it is difficult to establish cause and effect. To do so, an approximate analytical theory is needed.

The basis for this theory is the linearization of the 6 DOF equations of motion about a reference path which is characterized by constant angle of attack, Mach number, and elevator angle. Then, it is only necessary to solve linear ordinary differential equations with constant coefficients. From these results comes information about response to control input and stability.

First, the nonlinear 6 DOF equations for flight in a vertical plane (longitudinal motion) are stated, and valid approximations are introduced. Second, the resulting equations are linearized about a reference path which is a quasi-steady climb, cruise or descent. Third, the response of an airplane to an elevator input or a vertical gust is examined, and the stability characteristics are determined. The effect of cg change is discussed briefly, as is lateral-directional stability and control.

11.1   Equations of Motion

In Chap. 10, the 6 DOF equations of motion for flight in a vertical
plane have been derived for the regular body axes and for the stability
body axes. The latter is a set of body axes which is oriented at an
angle ϕ relative to the regular body axes. Dynamic stability and control
studies involve two flight paths (Fig. 11.1): a *reference path* and a
perturbed path which lies in the neighborhood of the reference path (small
perturbations).

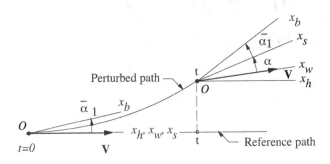

Figure 11.1: Perturbed Path and Reference Path

The reference path is assumed to be a steady cruise, climb, or
descent. It is possible to use the term steady because the time period of
interest for dynamic stability and control studies is sufficiently small that
atmospheric properties and mass properties can be assumed constant.
As a consequence, the angle of attack, the elevator angle, and the Mach
number are constant, and the pitch rate and the angle of attack rate are
zero on the reference path.

At $t = 0$, the airplane is on the reference path and is disturbed
from it, say by an elevator input. At this point, stability axes are defined
to be body axes which are aligned with the wind axes at $t = 0$. Hence,
the angle ϕ is the *reference angle of attack*, that is, $\phi = \bar{\alpha}_1$, so that
the stability axes change with the flight condition. For longitudinal
motion, the moment of inertia about the y_b axis doe not change with
flight condition.

In stability and control, it is the convention to use α and δ_E
to denote the changes (perturbations) in the angle of attack and the
elevator angle from the reference values. To do this, the angle of attack

and elevator angle are now called $\bar{\alpha}$ and $\bar{\delta}_E$. Hence, the angle of attack is written as $\bar{\alpha} = \bar{\alpha}_1 + \alpha$ where $\bar{\alpha}_1$ is the angle of attack on the reference path and α is the angle of attack perturbation. Similarly, $\bar{\delta}_E = \bar{\delta}_{E_1} + \delta_E$.

With $\phi = \bar{\alpha}_1$, the 6 DOF equations of motion in stability axes are obtained from Eqs. (10.20) as

$$
\begin{aligned}
\dot{x} &= U\cos\Theta + W\sin\Theta \\
\dot{h} &= U\sin\Theta - W\cos\Theta \\
\dot{U} &= -WQ + (1/m)[\, T\cos(\varepsilon_0 + \bar{\alpha}_1) \\
&\quad + L\sin\alpha - D\cos\alpha - mg\sin\Theta] \\
\dot{W} &= UQ + (1/m)[-T\sin(\varepsilon_0 + \bar{\alpha}_1) \\
&\quad - L\cos\alpha - D\sin\alpha + mg\cos\Theta] \\
\dot{\Theta} &= Q \\
\dot{Q} &= (M^A + M^T)/I_{yy} \\
\dot{m} &= -CT/g
\end{aligned}
\tag{11.1}
$$

where the airplane velocity and angle of attack are related to the velocity components as

$$
V = \sqrt{U^2 + W^2}, \quad \tan\alpha = W/U. \tag{11.2}
$$

It is important to recall that U, W are the components of the velocity vector on the stability axes, Θ is the pitch angle of the x_s axis, and α is the angle of attack of the x_s axis.

The functional relations for the forces and moments are

$$
\begin{aligned}
D = D(h, V, \bar{\alpha}), \quad L &= L(h, V, \bar{\alpha}, \bar{\delta}_E, Q, \dot{\bar{\alpha}}), \quad C = C(h, V, T) \\
M^A = M^A(h, V, \bar{\alpha}, \bar{\delta}_E, Q, \dot{\bar{\alpha}}), \quad M^T &= M^T(h, V, T).
\end{aligned}
\tag{11.3}
$$

Note that the thrust has been made a variable instead of the power setting.

Several approximations are made at this point. Since the time span of interest during any trajectory is small, the atmospheric properties of density and speed of sound and the properties of mass, center of mass, and moment of inertia are assumed constant. Since the translational kinematic equations are linked to the remaining equations through the atmospheric properties, they uncouple from the system and can be used later to compute the geometry of the perturbed trajectory, if needed. Also, the mass equation uncouples from the system and can be used later to compute the mass change. Other assumptions which are made are that α, Θ, and $\varepsilon_0 + \bar{\alpha}_1$ are small, $T(\varepsilon_0 + \bar{\alpha}_1) << mg$, and the thrust is constant. With these assumptions, the remaining equations of motion become

$$\dot{U} = -WQ + (1/m)(T + L\alpha - D - mg\Theta) \qquad (11.4)$$
$$\dot{W} = UQ + (1/m)(-L - D\alpha + mg) \qquad (11.5)$$
$$\dot{\Theta} = Q \qquad (11.6)$$
$$\dot{Q} = (M^A + M^T)/I_{yy} \qquad (11.7)$$

where Eqs. (11.2) with $\tan\alpha = \alpha$ and (11.3) still hold.

The equations for reference path (subscript 1) are obtained by making the substitutions

$$\dot{U}_1 = \dot{W}_1 = \dot{\Theta}_1 = \dot{Q}_1 = 0 \qquad (11.8)$$

which imply that U_1, W_1, Θ_1, and Q_1 are constant. Because stability axes are used, the x_s axis is along the x_w axis so that

$$W_1 = 0, \quad V_1 = U_1 \quad \alpha_1 = 0. \qquad (11.9)$$

The equations of motion become

$$T_1 - D_1 - mg\Theta_1 = 0 \qquad (11.10)$$
$$-L_1 + mg = 0 \qquad (11.11)$$
$$Q_1 = 0 \qquad (11.12)$$
$$M_1^A + M_1^T = 0 \qquad (11.13)$$

with

$$D = D(h, V, \bar{\alpha}), \quad L = L(h, V, \bar{\alpha}, \bar{\delta}_E)$$
$$M^A = M^A(h, V, \bar{\alpha}, \bar{\delta}_E), \quad M^T = M^T(h, V, T). \qquad (11.14)$$

Given $h_1, V_1, m_1 g$, and T_1, these equations can be solved for $\Theta_1 (= \gamma_1), \bar{\alpha}_1$, and $\bar{\delta}_{E_1}$ as in Chap. 9.

11.2   Linearized Equations of Motion

In order to linearize the thrust/aerodynamic terms separately, Eqs. (11.4) through (11.7) are rewritten as

$$\dot{U} = -WQ + (1/m)(C_x \bar{q} S - mg\Theta) \qquad (11.15)$$

$$\dot{W} = UQ + (1/m)(C_z \bar{q} S + mg) \tag{11.16}$$
$$\dot{\Theta} = Q \tag{11.17}$$
$$\dot{Q} = C_m \bar{q} S \bar{c} / I_{yy} \tag{11.18}$$

where S and $\bar{c}$ are the wing planform area and the wing mean aerodynamic chord. If $C_T \triangleq T/\bar{q}S$, the force and moment coefficients are given by

$$C_x = C_T + C_L \alpha - C_D$$
$$C_z = -C_L - C_D \alpha \tag{11.19}$$
$$C_m = C_m^T + C_m^A.$$

The equations for the reference path become

$$C_{x_1} \bar{q}_1 S - mg\Theta_1 = 0$$
$$C_{z_1} \bar{q}_1 S + mg = 0$$
$$Q_1 = 0 \tag{11.20}$$
$$C_{m_1} = 0$$

where for $\alpha_1 = 0$

$$C_{x_1} = C_{T_1} - C_{D_1}$$
$$C_{z_1} = -C_{L_1} \tag{11.21}$$
$$C_{m_1} = C_{m_1}^T + C_{m_1}^A.$$

The equations of motion are now linearized about the reference path. This is done by writing each variable at time t as the value on the reference path plus a small perturbation, that is,

$$U = U_1 + u$$
$$W = W_1 + w$$
$$\Theta = \Theta_1 + \theta \tag{11.22}$$
$$Q = Q_1 + q$$

where U_1 and Θ_1 are constant, $W_1 = 0$ in stability axes and $Q_1 = 0$ on the reference path. Also, the force and moment coefficients are written as

$$C_x = C_{x_1} + \Delta C_x$$
$$C_z = C_{z_1} + \Delta C_z \tag{11.23}$$
$$C_m = C_{m_1} + \Delta C_m$$

where $\Delta(\ )$ denotes a small quantity.

These definitions are substituted into the equations of motion and products and squares of small quantities (second-order terms) are neglected. First, consider Eq. (11.2) for the velocity, that is,

$$
\begin{aligned}
V &= \sqrt{U^2 + W^2} = \sqrt{(U_1 + u)^2 + w^2} \\
&= U_1 \left(1 + 2\tfrac{u}{U_1} + \tfrac{u^2}{U_1^2} + \tfrac{w^2}{U_1^2}\right)^{1/2}.
\end{aligned}
\tag{11.24}
$$

After applying the binomial expansion

$$
(1+x)^n = 1 + nx + \cdots
\tag{11.25}
$$

and neglecting second-order terms, it is seen that

$$
V = U_1 \left(1 + \frac{u}{U_1}\right) = U_1 + u.
\tag{11.26}
$$

Next, the equation (11.2) for the angle of attack perturbation leads to

$$
\alpha = \frac{W}{U} = \frac{w}{U_1 + u} = \frac{w}{U_1}\left(1 + \frac{u}{U_1}\right)^{-1} = \frac{w}{U_1}\left(1 - \frac{u}{U_1} + \cdots\right).
\tag{11.27}
$$

After neglecting second-order terms, it is seen that

$$
\alpha = \frac{w}{U_1}
\tag{11.28}
$$

Now, Eq. (11.15) for U can be written in terms of small quantities as

$$
\begin{aligned}
\tfrac{d}{dt}(U_1 + u) &= -wq + \tfrac{1}{m}\left[(C_{x_1} + \Delta C_x)\tfrac{1}{2}\rho(U_1 + u)^2 S \right. \\
&\quad \left. - \; mg(\Theta_1 + \theta)\right].
\end{aligned}
\tag{11.29}
$$

After neglecting second-order terms and accounting for the reference equation

$$
C_{x_1}\bar{q}_1 S - mg\Theta_1 = 0,
\tag{11.30}
$$

the equation for u becomes

$$
\dot{u} = \frac{1}{m}\left[\left(\Delta C_x + 2C_{x_1}\frac{u}{U_1}\right)\bar{q}_1 S - mg\theta\right]
\tag{11.31}
$$

If this process is applied to all of Eqs. (11.15) through (11.18), the small perturbation equations of motion become

$$
\begin{aligned}
\dot{u} &= (\Delta C_x + 2C_{x_1}\tfrac{u}{U_1})(\bar{q}_1 S/m) - g\theta \\
U_1\dot{\alpha} &= U_1 q + (\Delta C_z + 2C_{z_1}\tfrac{u}{U_1})(\bar{q}_1 S/m) \\
\dot{\theta} &= q \\
\dot{q} &= \Delta C_m \bar{q} S\bar{c}/I_{yy}.
\end{aligned}
\qquad (11.32)
$$

Note that Eq. (11.28) has been used to replace w by $U_1\alpha$ and that each term is linear in a small quantity.

The nondimensional forms of the functional relations (11.3) are given by

$$
\begin{aligned}
C_D &= C_D(\bar{\alpha}, M), \quad C_L = C_L(\bar{\alpha}, \bar{\delta}_E, M, \tfrac{\bar{c}Q}{2V}, \tfrac{\bar{c}\dot{\alpha}}{2V}) \\
C_m^T &= C_m^T(M, C_T), \quad C_m^A = C_m^A(\bar{\alpha}, \bar{\delta}_E, M, \tfrac{\bar{c}Q}{2V}, \tfrac{\bar{c}\dot{\alpha}}{2V}).
\end{aligned}
\qquad (11.33)
$$

In view of these functional forms, each force and moment coefficient C_x, C_z, or C_m has the same functional form. For example,

$$
C_x = C_x\left(\bar{\alpha}, \bar{\delta}_E, M, \frac{\bar{c}Q}{2V}, \frac{\bar{c}\dot{\alpha}}{2V}\right).
\qquad (11.34)
$$

Writing each argument as the reference value plus a small perturbation yields the relations

$$
\begin{aligned}
\bar{\alpha} &= \bar{\alpha}_1 + \alpha \\
\bar{\delta}_E &= \bar{\delta}_{E_1} + \delta_E \\
M &= \tfrac{U_1+u}{a} = M_1 + M_1\tfrac{u}{U_1} \\
\tfrac{\bar{c}Q}{2V} &= \tfrac{\bar{c}q}{2V} = \tfrac{\bar{c}q}{2U_1} \\
\tfrac{\bar{c}\dot{\alpha}}{2V} &= \tfrac{\bar{c}\dot{\alpha}}{2V} = \tfrac{\bar{c}\dot{\alpha}}{2U_1}.
\end{aligned}
\qquad (11.35)
$$

Now, C_x can be expanded in a Taylor series about the reference values as follows:

$$
\begin{aligned}
C_x &= C_{x_1} + \left.\frac{\partial C_x}{\partial \bar{\alpha}}\right|_1 \alpha + \left.\frac{\partial C_x}{\partial \bar{\delta}_E}\right|_1 \delta_E + \left.\frac{\partial C_x}{\partial M}\right|_1 M_1\tfrac{u}{U_1} \\
&+ \left.\frac{\partial C_x}{\partial \frac{\bar{c}Q}{2U_1}}\right|_1 \frac{\bar{c}q}{2U_1} + \left.\frac{\partial C_x}{\partial \frac{\bar{c}\dot{\alpha}}{2U_1}}\right|_1 \frac{\bar{c}\dot{\alpha}}{2U_1}.
\end{aligned}
\qquad (11.36)
$$

It is shown in Eq. (11.43) that $\dot{\alpha} = \dot{\alpha}(\alpha, \delta_E, u/U_1, \bar{c}q/2U_1)$ so that it is not really possible to take a derivative with respect to $\dot{\alpha}$ while holding the other variables constant. However, because the resulting equations are linear, it does not matter whether this functional relation is included now or whether $\dot{\alpha}$ is substituted later.

Since the reference values are all constant, each derivative can be expressed in terms of the nondimensional perturbation. For example,

$$\partial \bar{\alpha} = \partial \alpha \tag{11.37}$$

so that

$$\left.\frac{\partial C_x}{\partial \bar{\alpha}}\right|_1 = \left.\frac{\partial C_x}{\partial \alpha}\right|_1 . \tag{11.38}$$

Doing this for every variable and recalling that $C_x = C_{x_1} + \Delta C_x$ leads to

$$
\begin{aligned}
\Delta C_x &= \left.\frac{\partial C_x}{\partial \alpha}\right|_1 \alpha + \left.\frac{\partial C_x}{\partial \delta_E}\right|_1 \delta_E + \left.\frac{\partial C_x}{\partial \frac{u}{U_1}}\right|_1 \frac{u}{U_1} \\
&+ \left.\frac{\partial C_x}{\partial \frac{\bar{c}q}{2U_1}}\right|_1 \frac{\bar{c}q}{2U_1} + \left.\frac{\partial C_x}{\partial \frac{\bar{c}\dot{\alpha}}{2U_1}}\right|_1 \frac{\bar{c}\dot{\alpha}}{2U_1}
\end{aligned}
\tag{11.39}
$$

with identical expressions for ΔC_z and ΔC_m.

At this point, the following definitions are introduced:

$$
\begin{aligned}
C_{x_\alpha} &= \left.\frac{\partial C_x}{\partial \alpha}\right|_1, \quad C_{x_{\delta_E}} = \left.\frac{\partial C_x}{\partial \delta_E}\right|_1, \quad C_{x_u} = \left.\frac{\partial C_x}{\partial \frac{u}{U_1}}\right|_1 \\
C_{x_q} &= \left.\frac{\partial C_x}{\partial \frac{\bar{c}q}{2U_1}}\right|_1, \quad C_{x_{\dot{\alpha}}} = \left.\frac{\partial C_x}{\partial \frac{\bar{c}\dot{\alpha}}{2U_1}}\right|_1 .
\end{aligned}
\tag{11.40}
$$

With similar definitions for the C_z and C_m derivatives, the equations for the force and moment perturbations can be written as

$$
\begin{bmatrix} \Delta C_x \\ \Delta C_z \\ \Delta C_m \end{bmatrix} = \begin{bmatrix} C_{x_\alpha} & C_{x_{\delta_E}} & C_{x_u} & C_{x_q} & C_{x_{\dot{\alpha}}} \\ C_{z_\alpha} & C_{z_{\delta_E}} & C_{z_u} & C_{z_q} & C_{z_{\dot{\alpha}}} \\ C_{m_\alpha} & C_{m_{\delta_E}} & C_{m_u} & C_{m_q} & C_{m_{\dot{\alpha}}} \end{bmatrix} \begin{bmatrix} \alpha \\ \delta_E \\ \frac{u}{U_1} \\ \frac{\bar{c}q}{2U_1} \\ \frac{\bar{c}\dot{\alpha}}{2U_1} \end{bmatrix} \tag{11.41}
$$

All of the derivatives are evaluated on the reference path and are called *nondimensional stability derivatives*.

By combining Eqs. (11.32) and (11.41), the *linearized equations* or *small perturbation equations* become the following:

$$\dot{u} = \left[C_{x_\alpha} \alpha + C_{x_{\delta_E}} \delta_E + (C_{x_u} + 2C_{x_1}) \frac{u}{U_1} \right. \tag{11.42}$$

$$\left. + C_{x_q} \frac{\bar{c} q}{2U_1} + C_{x_{\dot\alpha}} \frac{\bar{c} \dot\alpha}{2U_1} \right] \frac{\bar{q}_1 S}{m} - g\theta$$

$$U_1 \dot{\alpha} = U_1 q + \left[C_{z_\alpha} \alpha + C_{z_{\delta_E}} \delta_E + (C_{z_u} + 2C_{z_1}) \frac{u}{U_1} \right. \tag{11.43}$$

$$\left. + C_{z_q} \frac{\bar{c} q}{2U_1} + C_{z_{\dot\alpha}} \frac{\bar{c} \dot\alpha}{2U_1} \right] \frac{\bar{q}_1 S}{m}$$

$$\dot{\theta} = q \tag{11.44}$$

$$\dot{q} = \left[C_{m_\alpha} \alpha + C_{m_{\delta_E}} \delta_E + (C_{m_u} + 2C_{m_1}) \frac{u}{U_1} \right. $$

$$\left. + C_{m_q} \frac{\bar{c} q}{2U_1} + C_{m_{\dot\alpha}} \frac{\bar{c} \dot\alpha}{2U_1} \right] \frac{\bar{q}_1 S \bar{c}}{I_{yy}} \tag{11.45}$$

The stability derivatives are evaluated on the reference path. Hence, they are all constant, as are C_{x_1}, C_{z_1}, and C_{m_1}. If Eq. (11.43) is solved for $\dot{\alpha}$ and if $\dot{\alpha}$ is eliminated from Eqs. (11.42) and (11.45), the result is a set of linear ordinary differential equations with constant coefficients. This set of equations has the general matrix form

$$\dot{x} = Fx + Gu \tag{11.46}$$

where the state and the control are defined as

$$x = [u\ \alpha\ \theta\ q]^T, \quad u = [\delta_E]. \tag{11.47}$$

The matrices F and G are the coefficient matrices of the system. This form of the equations is useful for automatic control studies.

The last step is to relate the force and moment coefficient derivatives to the usual thrust and aerodynamic coefficients. This is done by using Eqs. (11.19) and the definitions (11.40). To illustrate the process, consider the derivative C_{x_α}. By definition,

$$C_{x_\alpha} = \left. \frac{\partial C_x}{\partial \alpha} \right|_1 . \tag{11.48}$$

The coefficient C_x is related to the usual coefficients as (Eq. (11.19))

$$C_x = C_T + C_L \alpha - C_D \tag{11.49}$$

so that

$$C_{x_\alpha} = \left[\frac{\partial C_L}{\partial \alpha}\alpha + C_L - \frac{\partial C_D}{\partial \alpha}\right]_1 . \qquad (11.50)$$

Since $\alpha_1 = 0$,

$$C_{x_\alpha} = C_{L_1} - \left.\frac{\partial C_D}{\partial \alpha}\right|_1 = C_{L_1} - C_{D_\alpha}. \qquad (11.51)$$

Doing this for all derivatives leads to the following results:
C_x Derivatives:

$$\begin{aligned}
C_{x_\alpha} &= C_{L_1} - C_{D_\alpha} \\
C_{x_{\delta_E}} &= 0 \\
C_{x_u} &= C_{T_u} - C_{D_u} \qquad (11.52) \\
C_{x_q} &= 0 \\
C_{x_{\dot\alpha}} &= 0
\end{aligned}$$

C_z Derivatives:

$$\begin{aligned}
C_{z_\alpha} &= -C_{D_1} - C_{L_\alpha} \\
C_{z_{\delta_E}} &= -C_{L_{\delta_E}} \\
C_{z_u} &= -C_{L_u} \qquad (11.53) \\
C_{z_q} &= -C_{L_q} \\
C_{z_{\dot\alpha}} &= -C_{L_{\dot\alpha}}
\end{aligned}$$

C_m Derivatives:

$$\begin{aligned}
C_{m_\alpha} &= C_{m_\alpha}^A \\
C_{m_{\delta_E}} &= C_{m_{\delta_E}}^A \\
C_{m_u} &= C_{m_u}^A + C_{m_u}^T \qquad (11.54) \\
C_{m_q} &= C_{m_q}^A \\
C_{m_{\dot\alpha}} &= C_{m_{\dot\alpha}}^A .
\end{aligned}$$

With the exception of the u derivatives, formulas for calculating the stability derivatives have already been derived in Chap. 8. Because the nomenclature here is different from that in Chap. 8, the relationships between the two sets of nomenclature are presented in App. B, along with formulas for the u derivatives.

Values for the nondimensional stability derivatives have been calculated for the SBJ at $M = .6$. The results are listed in Sec. A.5.

11.3 Longitudinal Stability and Control

In the next two sections, the longitudinal response of an airplane to an elevator step input and to a vertical gust are determined. First, the small perturbation equations of motion are rewritten in a dimensional form, and Laplace transforms are used to find the response. Next, the response is examined to determine its nature and, hence, the longitudinal stability characteristics of the airplane. Then, the results are applied to the subsonic business jet of App. A. Finally, approximate modes are found, and the effect of cg position is examined.

The first step is to write the equations of motion (11.42) through (11.45) in a cleaner form. This is done by introducing the following *dimensional stability derivatives*:

$$
\begin{aligned}
X_\alpha &= \frac{\bar{q}_1 S C_{x\alpha}}{m} \; \frac{ft}{s^2} & X_u &= \frac{\bar{q}_1 S (C_{xu}+2C_{x_1})}{mU_1} \; \frac{1}{s} \\[2mm]
X_{\dot{\alpha}} &= \frac{\bar{q}_1 S \bar{c} C_{x\dot{\alpha}}}{2mU_1} \; \frac{ft}{s} & X_q &= \frac{\bar{q}_1 S \bar{c} C_{xq}}{2mU_1} \; \frac{ft}{s} \\[2mm]
X_{\delta_E} &= \frac{\bar{q}_1 S C_{x\delta_E}}{m} \; \frac{ft}{s^2} \\[2mm]
Z_\alpha &= \frac{\bar{q}_1 S C_{z\alpha}}{m} \; \frac{ft}{s^2} & Z_u &= \frac{\bar{q}_1 S (C_{zu}+2C_{z_1})}{mU_1} \; \frac{1}{s} \\[2mm]
Z_{\dot{\alpha}} &= \frac{\bar{q}_1 S \bar{c} C_{z\dot{\alpha}}}{2mU_1} \; \frac{ft}{s} & Z_q &= \frac{\bar{q}_1 S \bar{c} C_{zq}}{2mU_1} \; \frac{ft}{s} & (11.55) \\[2mm]
Z_{\delta_E} &= \frac{\bar{q}_1 S C_{z\delta_E}}{m} \; \frac{ft}{s^2} \\[2mm]
M_\alpha &= \frac{\bar{q}_1 S \bar{c} C_{m\alpha}}{I_{yy}} \; \frac{1}{s^2} & M_u &= \frac{\bar{q}_1 S \bar{c} (C_{mu}+2C_{m_1})}{I_{yy}U_1} \; \frac{1}{ft\,s} \\[2mm]
M_{\dot{\alpha}} &= \frac{\bar{q}_1 S \bar{c}^2 C_{m\dot{\alpha}}}{2I_{yy}U_1} \; \frac{1}{s} & M_q &= \frac{\bar{q}_1 S \bar{c}^2 C_{mq}}{2I_{yy}U_1} \; \frac{1}{s} \\[2mm]
M_{\delta_E} &= \frac{\bar{q}_1 S \bar{c} C_{m\delta_E}}{I_{yy}} \; \frac{1}{s^2}.
\end{aligned}
$$

Because the effects of $\dot{\alpha}$, q, and δ_E on the drag and the thrust have been neglected, it is seen from Eq. (11.52) that $X_{\dot{\alpha}}$, X_q, and X_{δ_E} are zero. Hence, after the pitch rate perturbation q is replaced by $\dot{\theta}$, Eqs. (11.42), (11.43), and (11.45) become

$$
\begin{aligned}
\dot{u} &= X_u u + X_\alpha \alpha - g\theta & (11.56) \\
U_1 \dot{\alpha} &= U_1 \dot{\theta} + Z_u u + Z_\alpha \alpha + Z_{\dot{\alpha}} \dot{\alpha} + Z_q \dot{\theta} + Z_{\delta_E} \delta_E & (11.57) \\
\ddot{\theta} &= M_u u + M_\alpha \alpha + M_{\dot{\alpha}} \dot{\alpha} + M_q \dot{\theta} + M_{\delta_E} \delta_E & (11.58)
\end{aligned}
$$

where the δ_E terms have been listed last because it is the control input.

Values for the dimensional stability derivatives have been calculated for the SBJ at $M = .6$. They are listed in Sec. A.5.

11.4   Response to an Elevator Step Input

Eqs. (11.56) through (11.58) are solved by using Laplace transforms which are reviewed briefly in App. C. The Laplace transform of these equations with zero initial conditions leads to

$$A(s) \begin{bmatrix} u(s) \\ \alpha(s) \\ \theta(s) \end{bmatrix} = b\delta_E(s) \tag{11.59}$$

where

$$A(s) = \begin{bmatrix} s - X_u & -X_\alpha & g \\ -Z_u & (U_1 - Z_{\dot\alpha})s - Z_\alpha & -(U_1 + Z_q)s \\ -M_u & -(M_\alpha + M_{\dot\alpha}s) & s^2 - M_q s \end{bmatrix} \quad b = \begin{bmatrix} 0 \\ Z_{\delta_E} \\ M_{\delta_E} \end{bmatrix} \tag{11.60}$$

Eqs. (11.59) can be written in the form

$$\begin{bmatrix} A_{11} & A_{12} & A_{13} \\ A_{21} & A_{22} & A_{23} \\ A_{31} & A_{32} & A_{33} \end{bmatrix} \begin{bmatrix} u(s)/\delta_E(s) \\ \alpha(s)/\delta_E(s) \\ \theta(s)/\delta_E(s) \end{bmatrix} = \begin{bmatrix} b_1 \\ b_2 \\ b_3 \end{bmatrix} \tag{11.61}$$

The solution of this equation for $u(s)/\delta_E(s)$ follows from Cramer's rule to be

$$\frac{u(s)}{\delta_E(s)} = \frac{\begin{vmatrix} b_1 & A_{12} & A_{13} \\ b_2 & A_{22} & A_{23} \\ b_3 & A_{32} & A_{33} \end{vmatrix}}{\begin{vmatrix} A_{11} & A_{12} & A_{13} \\ A_{21} & A_{22} & A_{23} \\ A_{31} & A_{32} & A_{33} \end{vmatrix}}. \tag{11.62}$$

Similar expressions hold for $\alpha(s)/\delta_E(s)$ and $\theta(s)/\delta_E(s)$. Note that the denominator of each solution is the determinant of the matrix A or det A.

For a step input $\delta_E(s) = \delta_E/s$ where δ_E is constant, the $u(s)$ response is given by

$$u(s) = \frac{\delta_E}{s} \frac{\begin{vmatrix} b_1 & A_{12} & A_{13} \\ b_2 & A_{22} & A_{23} \\ b_3 & A_{32} & A_{33} \end{vmatrix}}{\det A}. \qquad (11.63)$$

To use partial fractions to return to the time domain, the roots of the denominator are determined from

$$\det A = 0. \qquad (11.64)$$

For system matrix A in (11.60), this equation leads to the fourth power polynomial

$$as^4 + bs^3 + cs^2 + ds + e = 0 \qquad (11.65)$$

where

$$
\begin{aligned}
a &= U_1 - Z_{\dot{\alpha}} \\
b &= -[(U_1 - Z_{\dot{\alpha}})M_q + Z_\alpha + M_{\dot{\alpha}}(U_1 + Z_q) - X_u(U_1 - Z_{\dot{\alpha}}) \\
c &= Z_\alpha M_q - M_\alpha(U_1 + Z_q) + X_u[(U_1 - Z_{\dot{\alpha}})M_q + Z_\alpha \\
  &\quad + M_{\dot{\alpha}}(U_1 + Z_q)] - X_\alpha Z_u \\
d &= -X_u[Z_\alpha M_q - M_\alpha(U_1 + Z_q)] + X_\alpha[Z_u M_q \\
  &\quad - M_u(U_1 + Z_q)] + g[Z_u M_{\dot{\alpha}} + M_u(U_1 - Z_{\dot{\alpha}})] \\
e &= g[Z_u M_\alpha - M_u Z_\alpha]
\end{aligned}
\qquad (11.66)
$$

The *characteristic equation* (11.65) admits four solutions (λ_1, λ_2, λ_3, λ_4) which may be real, complex, or a combination. The expanded form of Eq. (11.63) has the form

$$u(s) = \frac{u_s}{s} + \frac{u_1}{s - \lambda_1} + \frac{u_2}{s - \lambda_2} + \frac{u_3}{s - \lambda_3} + \frac{u_4}{s - \lambda_4} \qquad (11.67)$$

where the constants u_s, u_1 etc., are functions of the λ's. If the roots are unique, the *response* in the time domain is given by

$$
\begin{aligned}
u &= u_s + u_1 e^{\lambda_1 t} + u_2 e^{\lambda_2 t} + u_3 e^{\lambda_3 t} + u_4 e^{\lambda_4 t} & (11.68) \\
\alpha &= \alpha_s + \alpha_1 e^{\lambda_1 t} + \alpha_2 e^{\lambda_2 t} + \alpha_3 e^{\lambda_3 t} + \alpha_4 e^{\lambda_4 t} & (11.69) \\
\theta &= \theta_s + \theta_1 e^{\lambda_1 t} + \theta_2 e^{\lambda_2 t} + \theta_3 e^{\lambda_3 t} + \theta_4 e^{\lambda_4 t} & (11.70)
\end{aligned}
$$

These equations give the shape of the response at a single flight condition, that is, altitude, velocity, mass, thrust, center of gravity, moment of inertia about the pitch axis, and elevator deflection.

The stability of the response is determined from the transient or time dependent part of these equations. It is seen that real roots must be negative and that the real parts of complex roots must be negative to prevent the disturbances u, α, and θ from growing with time, that is, to have a dynamically stable airplane. Hence, airplane stability is determined by the roots of the characteristic equation det $A=0$.

Note that the total motion is the sum of several motions called *modes*. A real root corresponds to a nonoscillatory motion as illustrated in Fig. 11.2. A non-oscillatory mode is characterized by a single parameter $T = -1/\lambda$ called the *time constant*. A complex pair $\lambda_{p,q} = n \pm i\omega$ forms an oscillatory motion such as that shown in Fig. 11.3. An oscillatory mode, because it is created by two λ's, is characterized by two parameters: the *natural frequency* ω_n and the *damping ratio* ζ. These parameters can be obtained from the location of the roots (shown in Fig. 11.4) using the formulas

$$\begin{aligned} \omega_n &= |\lambda| = \sqrt{n^2 + \omega^2} \\ \zeta &= \cos\theta = \frac{-n}{\omega_n} . \end{aligned}$$

(11.71)

For the SBJ at $M = .6$ and $\bar{X}_{cg} = .3$, the coefficients (11.66) have the values

$$\begin{aligned} a &= 598 \\ b &= 1,400 \\ c &= 9840 \\ d &= 128 \\ e &= 80.7 \end{aligned}$$

(11.72)

The characteristic equation admits two complex roots, that is,

$$\lambda_{1,2} = -1.16 \pm 3.88j, \quad \lambda_{3,4} = -.0059 \pm .0940j.$$

(11.73)

The first complex root represents a high-frequency, highly-damped oscillation called the *short-period mode*, and the second complex root represents a low-frequency, lightly-damped oscillation called the *phugoid mode*. Both modes are stable. The natural frequency and the damping

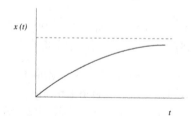

Figure 11.2: Example of a Stable Real Root Mode

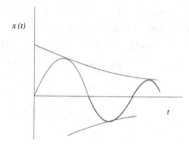

Figure 11.3: Example of a Stable Complex Root Mode

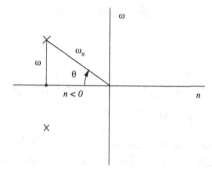

Figure 11.4: Damping Ratio and Natural Frequency

ratio for these modes are given by

$$
\begin{aligned}
\omega_{n_{sp}} &= 4.05 \text{ rad/s} & \omega_{n_p} &= .0906 \text{ rad/s} \\
\zeta_{sp} &= .287 & \zeta_p &= .0654
\end{aligned}
\tag{11.74}
$$

Note that the response has a steady part due to the deflection of the elevator, a part due to the short period mode which dies out rapidly, and a part due to the phugoid mode which dies out slowly. Because of the gentle character of the phugoid mode, it is the short-period mode that is actually controlled.

11.4.1 Approximate short-period mode

The short-period mode takes place over such a small time interval that the velocity is approximately constant ($u \cong 0$). Hence, an approximate result can be obtained by setting $u = 0$ and ignoring the $\dot{u}$ equation (11.56). The *approximate short-period mode* is governed by

$$
\begin{aligned}
U_1\dot{\alpha} - U_1\dot{\theta} &= Z_\alpha\alpha + Z_{\dot{\alpha}}\dot{\alpha} + Z_q\dot{\theta} + Z_{\delta_E}\delta_E & (11.75) \\
\ddot{\theta} &= M_\alpha\alpha + M_{\dot{\alpha}}\dot{\alpha} + M_q\dot{\theta} + M_{\delta_E}\delta_E & (11.76)
\end{aligned}
$$

so that $\det A = 0$ yields

$$
\begin{vmatrix}
(U_1 - Z_{\dot{\alpha}})s - Z_\alpha & -(U_1 + Z_q)s \\
-(M_\alpha + M_{\dot{\alpha}}s) & s^2 - M_q s
\end{vmatrix} = 0
\tag{11.77}
$$

Hence, the characteristic equation is the third-order polynomial

$$
s(as^2 + bs + c) = 0
\tag{11.78}
$$

where

$$
\begin{aligned}
a &= U_1 - Z_{\dot{\alpha}} \\
b &= -[(U_1 - Z_{\dot{\alpha}})M_q + Z_\alpha + M_{\dot{\alpha}}(U_1 + Z_q)] \\
c &= Z_\alpha M_q - M_\alpha(U_1 + Z_q)
\end{aligned}
\tag{11.79}
$$

The root $s = 0$ indicates neutral stability in pitch. Note that the approximate short-period is not affected by the derivative M_u.

For the SBJ at $M = .6$ and $\bar{X}_{cg} = .3$, the coefficients of the characteristic equation are given by

$$
a = 598, \quad b = 1,390, \quad c = 9,830.
\tag{11.80}
$$

Hence, the roots of the characteristic equation become

$$\lambda_{1,2} = -1.16 \pm 3.88j \qquad (11.81)$$

leading to the following natural frequency and damping ratio:

$$
\begin{aligned}
\omega_{n_{sp}} &= 4.05 \text{ rad/s} \\
\zeta_{sp} &= 0.287
\end{aligned}
\qquad (11.82)
$$

which compare well with the exact numbers (11.71).

11.4.2 Approximate phugoid mode

During the phugoid motion, the angle of attack is approximately constant. Hence, an approximate result can be obtained by setting $\alpha = 0$ and ignoring the moment equation (11.58). The equations of motion for the *approximate phugoid mode* are given by

$$
\begin{aligned}
\dot{u} &= +X_u u - g\theta & (11.83) \\
-U_1\dot{\theta} &= Z_u u + Z_q\dot{\theta} + Z_{\delta_E}\delta_E, & (11.84)
\end{aligned}
$$

Hence, the system matrix is given by

$$
\begin{vmatrix}
s - X_u & g \\
-Z_u & -(U_1 + Z_q)s
\end{vmatrix} = 0
\qquad (11.85)
$$

and yields the characteristic equation

$$as^2 + bs + c = 0 \qquad (11.86)$$

where

$$
\begin{aligned}
a &= -(U_1 + Z_q) \\
b &= X_u(U_1 + Z_q) \\
c &= Z_u g.
\end{aligned}
\qquad (11.87)
$$

Note that the approximate phugoid mode is not affected by the derivative M_u.

For the SBJ at $M = .6$ and $\bar{X}_{cg} = .3$ the coefficients of the characteristic equation are given by

$$a = -594, \quad b = -6.70, \quad c = -3.98. \qquad (11.88)$$

Hence, the roots of the characteristic equation become

$$\lambda_{1,2} = -.0056 \pm .0817j \qquad (11.89)$$

leading to the following natural frequency and damping ratio:

$$\begin{aligned} \omega_{n_p} &= .0818 \text{ rad/s} \\ \zeta_p &= .0689 \end{aligned} \qquad (11.90)$$

The approximate results for the phugoid mode differ from the exact results by about 10%. The approximation gets worse as the cg is moved aft toward the neutral point.

11.5 Response to a Gust

It is desired to find the response of an airplane to a step input in the vertical speed of the atmosphere. However, to derive the equations of motion, it is assumed that the vertical wind speed is an arbitrary function of time, that is, for $t \geq 0$

$$w_x = 0 \ (\dot{w}_x = 0), \quad w_h = w_h(x) \ (\dot{w}_h \neq 0) \qquad (11.91)$$

where $x = x(t)$.

The equations of motion for flight in a moving atmosphere in stability axes are given in Sec. 10.3 and are rewritten here as

$$\begin{aligned} \dot{U} &= -WQ + (1/m)[T\cos(\varepsilon_0 + \bar{\alpha}_1) + L\sin\alpha - D\cos\alpha \\ &\quad - -mg\sin\Theta] - \dot{w}_x\cos\Theta - \dot{w}_h\sin\Theta \\ \dot{W} &= UQ - (1/m)[T\sin(\varepsilon_0 + \bar{\alpha}_1) + L\cos\alpha + D\sin\alpha \\ &\quad - mg\cos\Theta] - \dot{w}_x\sin\Theta + \dot{w}_h\cos\Theta \\ \dot{\Theta} &= Q \\ \dot{Q} &= M/I_{yy} \end{aligned} \qquad (11.92)$$

Making the usual approximations (Sec. 11.1) and $\dot{w}_x = 0$ leads to

$$\begin{aligned} \dot{U} &= -WQ + (1/m)(C_x\bar{q}S - mg\Theta) - \dot{w}_h\Theta \\ \dot{W} &= UQ + (1/m)(C_z\bar{q}S + mg) + \dot{w}_h \\ \dot{\Theta} &= Q \\ \dot{Q} &= M/I_{yy} \end{aligned} \qquad (11.93)$$

where C_x, C_z and C_m are defined in Eq. 11.19.

Next, the linearization of these equations for small $\dot{w}_h$ leads to

$$\begin{aligned}
\dot{u} &= X_u u + X_\alpha \alpha - g\theta - \Theta_1 \dot{w}_h \\
U_1 \dot{\alpha} &= U_1 \dot{\theta} + Z_u u + Z_\alpha \alpha + Z_{\dot{\alpha}} \dot{\alpha} + Z_q \dot{\theta} + Z_{\delta_E} \delta_E + \dot{w}_h \\
\ddot{\theta} &= M_u u + M_\alpha \alpha + M_{\dot{\alpha}} \dot{\alpha} + M_q \dot{\theta} + M_{\delta_E} \delta_E.
\end{aligned} \qquad (11.94)$$

Taking the Laplace transform, assuming zero elevator input, and assuming zero initial conditions including $w_h(0) = 0$ gives

$$A(s) \begin{bmatrix} u(s) \\ \alpha(s) \\ \theta(s) \end{bmatrix} = csw_h(s) \qquad (11.95)$$

where A is defined in Eq. (11.60) and C is given by

$$c = \begin{bmatrix} -\Theta_1 \\ 1 \\ 0 \end{bmatrix}. \qquad (11.96)$$

Eqs. (11.95) can be written in the expanded form

$$\begin{bmatrix} A_{11} & A_{12} & A_{13} \\ A_{21} & A_{22} & A_{23} \\ A_{31} & A_{32} & A_{33} \end{bmatrix} \begin{bmatrix} u(s)/w_h(s) \\ \alpha(s)/w_h(s) \\ \theta(s)/w_h(s) \end{bmatrix} = \begin{bmatrix} c_1 \\ c_2 \\ c_3 \end{bmatrix} s \qquad (11.97)$$

The solution of this equation for $u(s)/w_h(s)$ follows from Cramer's rule to be

$$\frac{u(s)}{w_h(s)} = s \frac{\begin{vmatrix} c_1 & A_{12} & A_{13} \\ c_2 & A_{22} & A_{23} \\ c_3 & A_{32} & A_{33} \end{vmatrix}}{\begin{vmatrix} A_{11} & A_{12} & A_{13} \\ A_{21} & A_{22} & A_{23} \\ A_{31} & A_{32} & A_{33} \end{vmatrix}} \qquad (11.98)$$

Similar expressions hold for $\alpha(s)/w_h(s)$ and $\theta(s)/w_h(s)$. Note that the denominator of each solution is the determinant of the matrix A or det A.

Consider the response to a step input in the vertical speed of the atmosphere, that is, $w_h(s) = w_h/s$ where w_h is a constant. For the $u(s)$ response,

$$u(s) = w_h \frac{\begin{vmatrix} c_1 & A_{12} & A_{13} \\ c_2 & A_{22} & A_{23} \\ c_3 & A_{32} & A_{33} \end{vmatrix}}{\det A}. \qquad (11.99)$$

To use partial fractions, the roots of the denominator are determined from $\det A = 0$ which leads to the same characteristic equation (11.65). If partial fractions is performed for unique roots, the response in the time domain is given by

$$u = u_1 e^{\lambda_1 t} + u_2 e^{\lambda_2 t} + u_3 e^{\lambda_3 t} + u_4 e^{\lambda_4 t} \qquad (11.100)$$

$$\alpha = \alpha_1 e^{\lambda_1 t} + \alpha_2 e^{\lambda_2 t} + \alpha_3 e^{\lambda_3 t} + \alpha_4 e^{\lambda_4 t} \qquad (11.101)$$

$$\theta = +\theta_1 e^{\lambda_1 t} + \theta_2 e^{\lambda_2 t} + \theta_3 e^{\lambda_3 t} + \theta_4 e^{\lambda_4 t} \qquad (11.102)$$

where the constants u_1, u_2 etc., are functions of the λ's. These equations give the shape of the response at a single flight condition and gust speed. The stability and response discussion for this problem is the same as that in Sec. 11.4

11.6 CG Effects

The characteristics of the short period mode and the phugoid mode are affected by cg position, as shown in Tables 11.1 and 11.2 for the SBJ. As the cg moves aft from .150 $\bar{c}$ to .512 $\bar{c}$ (the aerodynamic center or neutral point), the short-period mode is stable, the complex root moves toward the real axis, and the complex root splits into two stable real roots. On the other hand, the phugoid mode is stable until the cg gets near the aerodynamic center where it becomes unstable. The root moves away from the real axis. For a range of cg positions, the root does not change appreciably. When the cg nears the neutral point, the phugoid mode becomes unstable. This is not a problem because the most aft allowable position is .35 $\bar{c}$ for the SBJ.

While it is not shown here, the approximate phugoid equations predict that the phugoid roots do not change with cg position. Hence, the approximate phugoid mode is not valid for all cg positions.

Table 11.1: Effect of cg Position on the SBJ Short-period Mode

$$\bar{X}_{ac} = .512$$

$\bar{X}_{cg}$	$\bar{X}_{ac} - \bar{X}_{cg}$	Roots	ω_n	ζ
.150	.3620	-1.23 ± 5.10 j	5.24	.235
.200	.3120	-1.21 ± 4.73 j	4.83	.243
.250	.2620	-1.19 ± 4.33 j	4.49	.264
.300	.2120	-1.16 ± 3.88 j	4.05	.287
.350	.1620	-1.14 ± 3.38 j	3.57	.320
.400	.1120	-1.12 ± 2.79 j	3.00	.372
.450	.0620	-1.10 ± 2.03 j	2.31	.476
.475	.0370	-1.09 ± 1.51 j	1.86	.584
.500	.0120	-1.08 ± .685 j	1.28	.845
.509	.0000	-1.71, -.503		

Table 11.2: Effect of cg Position on the SBJ Phugoid Mode

$$\bar{X}_{ac} = .512$$

$\bar{X}_{cg}$	$\bar{X}_{ac} - \bar{X}_{cg}$	Roots	ω_n	ζ
.150	.3620	-.0059 ± .0865 j	.0867	.0676
.200	.3120	-.0059 ± .0874 j	.0876	.0670
.250	.2620	-.0059 ± .0887 j	.0889	.0663
.300	.2120	-.0059 ± .0940 j	.0906	.0654
.350	.1620	-.0060 ± .0932 j	.0933	.0639
.400	.1120	-.0060 ± .0979 j	.0981	.0610
.450	.0620	-.0058 ± .1090 j	.1090	.0535
.475	.0370	-.0050 ± .1210 j	.1210	.0417
.500	.0120	.0016 ± .1540 j	.1540	-.0102
.509	.0000	.0285 ± .1940 j	.1960	-.1450

11.7   Dynamic Lateral-Directional S&C

If the complete set of six-degree-of-freedom equations of motion (3D flight) is linearized, it splits apart into two subsets. One subset is composed of the equations for longitudinal motion, and the other subset is composed of the equations for lateral-directional motion. The lateral-directional motion of an airplane is associated with the side force, the rolling moment, and the yawing moment, and its state variables are the sideslip angle perturbation, the roll rate perturbation, and the yaw rate perturbation, while its controls are the rudder angle perturbation and the aileron angle perturbation.

When the linearized lateral-directional equations are investigated, it is seen that three modes exist: a non-oscillatory mode called the *spiral mode* which is usually unstable, a non-oscillatory mode called the *roll mode* which is stable, and an oscillatory mode called the *dutch roll mode* which is stable.

For the SBJ, the following results have been obtained:

Spiral mode: divergent $T_S = 992$ s

Roll mode: convergent $T_R = 1.97$ s

Dutch roll mode: convergent $\omega_n = 1.63$ rad/s, $\zeta = 0.036$

Even through the spiral mode is unstable, the time constant is sufficiently large that corrective action is easily taken by the pilot. Note that while the dutch roll mode is stable, its damping is quite small.

Design problems associated with lateral-directional motion include making the spiral mode stable or at least less unstable and making the damping of the dutch roll mode larger.

Problems

11.1 Derive the response of a first-order system with a nonzero initial conditions to a step input. For a stable system, the initial-condition response always dies out in time.

11.2 Derive the response of a second-order system $(0 < \zeta < 1)$ with nonzero initial conditions to a step input. Show that for a stable system the initial-condition response dies out in time.

11.3 Determine the response of a first-order system with zero initial conditions to an impulse $c\,\delta(t)$ where c is a constant. Is the response stable? Derive the steady state response for a stable system using both the time response as $t \to \infty$ and the final value theorem.

11.4 Determine the response of a second-order system with zero initial conditions to an impulse $c\,\delta(t)$ where c is a constant. Is the response stable? Derive the steady state response for a stable system using both the time response as $t \to \infty$ and the final value theorem.

11.5 Consider the motion of a pendulum having a mass m at the end of a massless rod of length l which makes an angle Φ with the vertical. The mass is subjected to a force of magnitude F perpendicular to the rod.

    a. Show that the equation of motion is given by

$$\ddot{\Phi} + p^2 \sin \Phi = f$$

    where $p^2 = g/l$ and $f = F/ml$.

    b. Assume that the amplitude of the oscillation is small, that is, $\Phi = \Phi_1 + \phi$ where $\Phi_1 = 0$, and derive the linearized equation of motion.

    c. Is the system stable? Examine roots of characteristic equation.

    d. What is the response of the system if $F = 0$, $\phi(0) \neq 0$, and $\dot{\phi}(0) = 0$.

    e. What is the response of the system if $F = 0$, $\phi(0) = 0$, and $\dot{\phi}(0) \neq 0$.

    f. What is the response of the system if F is an impulse, $\phi(0) = 0$, and $\dot{\phi}(0) = 0$.

    g. What is the response of the system if F is a step, $\phi(0) = 0$ and $\dot{\phi}(0) = 0$.

11.6 The equation of motion and initial conditions for a jet-powered car are given by

$$\ddot{X} = (g/W)[T - \mu W - (1/2)C_D \rho S \dot{X}^2]$$

$$t_0 = 0, \quad X(t_0) = 0, \quad \dot{X}(t_0) = \text{Given}$$

where $g, T, \mu, W, C_D, \rho, S$ are constant. It is desired to find the speed response of the car traveling at constant speed ($\dot{X}_n =$ Const) to a small change in the thrust from a nominal value, that is, $T = T_n + \tau(t)$. The steady-state equation of motion of the car is given by

$$\dot{X}_n = \sqrt{\frac{T_n - \mu W}{(1/2)C_D\rho S}} = Const, \quad \ddot{X}_n = 0.$$

Assuming that $X = X_n + x$ where x is small, linearize the equation of motion about the nominal solution and show that

$$\ddot{x} = -\alpha \dot{x} + (g/W)\tau$$

where

$$\alpha = (g/W)C_D\rho S \sqrt{\frac{T_n - \mu W}{(1/2)C_D\rho S}}.$$

Also show that

$$x(t_0) = 0, \quad \dot{x}(t_0) = 0.$$

Show that the velocity response $\dot{x}(t)$ to a step input in the thrust ($\tau(s) = \tau/s$, where τ is constant) is given by

$$\dot{x}(t) = \frac{g\tau}{W\alpha}(1 - e^{-\alpha t})$$

and that it is stable.

11.7 For the flight condition of Sec. A.2, calculate the nondimensional and dimensional stability derivatives (App.B).

11.8 Using the dimensional stability derivatives in Sec. A.5, derive the coefficients of the characteristic equation (11.65) for the SBJ. What are the modes of the motion, and what are the constants governing each mode (T or ζ, ω_n)?

11.9 Using the dimensional stability derivatives in Sec. A.5, derive the coefficients of the characteristic equation for the approximate short-period mode. What are the modes of the motion, and what are the constants governing each mode (T or ζ, ω_n)?

11.10 Using the dimensional stability derivatives in Sec. A.5, derive the coefficients of the characteristic equation for the approximate phugoid mode. What are the modes of the motion, and what are the constants governing each mode (T or ζ, ω_n)?

11.11 It is not really possible to input an impulse in the elevator angle. However, it is possible to show that a short-time pulse has the same Laplace transform as the impulse. A pulse is a step input at $t = 0$ followed by a negative step input of the same magnitude at time $t = T$. Show that a short-time pulse has the approximate Laplace transform AT, where A is the magnitude of the pulse and T is its duration.

11.12 Consider an airplane in quasi-steady level flight, and assume that at some time instant, the airplane is disturbed from this motion by a small aileron input. The roll mode of an airplane can be studied by analyzing the equations

$$\dot{\phi} = p$$
$$\dot{p} = L_p p + L_{\delta_A} \delta_A$$

where ϕ is the roll angle perturbation, p is the roll rate perturbation, $L_p < 0$ is the constant roll damping derivative, $L_{\delta_A} < 0$ is the constant roll control derivative, and δ_A is the aileron deflection perturbation. Assume zero initial conditions.

a. What is the response of the airplane to an aileron pulse where $\delta_A(s) = \delta_A T$. See Prob. 11.11. Is the response stable? What is the steady state response?

b. What is the response of the airplane to an aileron step. Is the response stable? What is the steady state response?

Appendix A

SBJ Data and Calculations

The airplane used for all of the calculations in this text is the subsonic business jet (SBJ) shown in Fig. A.1. It is powered by two GE CJ610-6 turbojets. This airplane resembles an early model of the Lear Jet, but there are many differences. As an example, the airfoils have been chosen so that aerodynamic data is readily available. The maximum take-off weight is 13,000 lb which includes 800 lb of reserve fuel and 4,200 lb of climb/cruise fuel. Hence, the zero-fuel weight is 8,000 lb. For the calculations of this appendix, it is assumed that the center of gravity is located 21.4 ft from the nose. The cg range is from 20.4 ft from the nose to 21.8 ft from the nose, a total of 1.4 ft. In terms of the wing mean aerodynamic chord, $\bar{c}$, the cg varies between $.15\bar{c}$ and $.35\bar{c}$.

The geometric characteristics of the SBJ are developed in Sec. A.2. Basic dimensions are measured from the three-view drawing (Fig, A.1), and the remaining geometric quantities are calculated using formulas presented in the text (Chap. 3). Next, the flight condition which is used to make aerodynamic and stability and control calculations is stated in Sec. A.2. It is level flight ($\gamma=0$) at $h=30,000$ ft, $V = 597$ ft/s ($M=0.6$), and $W = 11,000$ lb ($m = 342$ slugs); the cg position is assumed to be 21.4 ft from the nose ($.30\bar{c}$)

In Sec. A.3, aerodynamics of the wing, the wing-body combination, the horizontal tail, and the whole airplane are calculated using the formulas of Chap. 8. In all calculations, the aerodynamics of the wing-body combination are approximated by those of the entire wing alone. Next, the trim angle of attack and the elevator deflection are calculated in Sec. A.4, using the formulas derived in Chap. 9. Finally, in

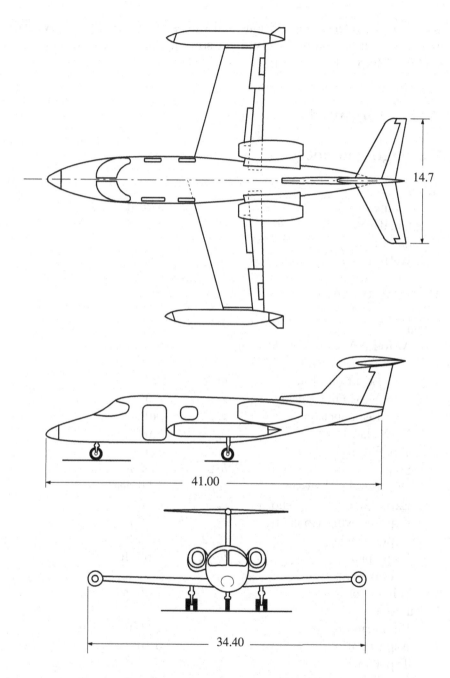

Figure A.1: Turbojet SBJ Three-view Drawing

Sec. A.5, the reference conditions and values of the stability derivatives needed for the dynamic longitudinal stability and control calculations made in Chap. 11 are given.

A.1   Geometry

Fuselage geometry

Given:

Diameter	d	$=$	5.25 ft
Length	l	$=$	41.0 ft

Calculated:

Wetted area	S_{wet}	$=$	456. ft^2

Wing geometry

Given:

Airfoil NACA 64-109 (M=0)

Thickness ratio	t/c	$=$	.09
Location of ac	ac/c	$=$	.258
Maximum thickness	$x_{mt}/c =$		.35
Peak suction	$x_{ps}/c =$		.40
Root chord	c_r	$=$	9.00 ft
Tip chord	c_t	$=$	4.50 ft
Span	$b/2$	$=$	17.2 ft
Quarter-chord sweep	Λ_{qc}	$=$	13.0 deg
Exposed wing root chord	c_{r_e}	$=$	8.05 ft
Exposed wing tip chord	c_{t_e}	$=$	4.90 ft
Exposed wing span	$b_e/2$	$=$	13.3 ft
Flap span	$b_F/2$	$=$	9.63 ft
Average flap chord	c_F	$=$	1.31 ft
Wing chord at c_F	c	$=$	7.35 ft

Calculated:

Planform area	S	$=$	232. ft^2
Aspect ratio	A	$=$	5.10
Taper ratio	λ	$=$	.500
Leading-edge sweep	Λ_{le}	$=$	16.5 deg
Maximum-thickness sweep	Λ_{mt}	$=$	11.6 deg
Peak-suction sweep	Λ_{ps}	$=$	10.8 deg

Half-chord sweep	Λ_{hc}	=	9.40 deg
Mean aerodynamic chord	$\bar{c}$	=	7.00 ft
y coordinate of ac	η	=	7.64 ft
x coordinate of ac	ξ	=	4.07 ft
Flap chord ratio	c_F/c	=	.178
Flap span ratio	b_F/b	=	.560
Body span ratio	d_B/b	=	.153
Root chord body length ratio	c_r/l_B	=	.220
Wetted area	S_{wet}	=	344. ft^2

Horizontal tail geometry

Given:

Airfoil NACA 64-008 (M=0)

Thickness ratio	t/c	=	.08
Location of ac	ac/c	=	.260
Root chord	c_r	=	5.00 ft
Tip chord	c_t	=	2.35 ft
Span	$b/2$	=	7.35 ft
Quarter-chord sweep	Λ_{qc}	=	25.0 deg
Average elevator chord	c_E	=	1.18 ft
Horizontal tail chord at c_E	c	=	3.74 ft

Calculated:

Planform area	S	=	54.0 ft^2
Aspect ratio	A	=	4.00
Taper ratio	λ	=	.470
Leading-edge sweep	Λ_{le}	=	29.1 deg
Half-chord sweep	Λ_{hc}	=	20.6 deg
Mean aerodynamic chord	$\bar{c}$	=	3.83 ft
y coordinate of ac	η	=	3.23 ft
x coordinate of ac	ξ	=	2.80 ft
Elevator chord ratio	c_E/c	=	0.316
Wetted area	S_{wet}	=	108. ft^2

Vertical tail geometry

Given:

Airfoil NACA 64-010

Thickness ratio	t/c	=	.10
Root chord	c_r	=	9.05 ft
Tip chord	c_t	=	4.19 ft
Span	b	=	5.70 ft
Quarter-chord sweep	Λ_{qc}	=	40.0 deg

Calculated:

Planform area	S	=	37.7 ft^2
Aspect ratio	A	=	1.72
Taper ratio	λ	=	.463
Mean aerodynamic chord	$\bar{c}$	=	6.92 ft
Wetted area	S_{wet}	=	75.4 ft^2

Nacelle geometry

Given:

Diameter	d	=	2.30 ft
Length	l	=	7.70 ft

Calculated:

Wetted area	S_{wet}	=	55.6 ft^2

Tip tank geometry

Given:

Diameter	d	=	1.75 ft
Length	l	=	14.0 ft

Calculated:

Wetted area	S_{wet}	=	61.2 ft^2

Airplane geometry

Given:

Wing incidence	i_W	=	1.00 deg
Wing dihedral	Γ_W	=	2.50 deg
Horizontal tail incidence	i_H	=	-2.0 deg
Horizontal tail dihedral	Γ_H	=	0.00 deg
Thrust moment arm	l_T	=	-2.0 ft

Nose to wing apex		= 17.0 ft
Nose to wing ac		= 21.1 ft
Nose to wing mac le		= 19.3 ft
Nose to HT apex		= 37.1 ft
Nose to HT ac		= 39.9 ft
Nose to HT mac le		= 38.9 ft
Wing ac to HT ac	l_H	= 18.8 ft
Wing ac to HT ac	h_H	= 7.88 ft

Calculated:

Wetted area	S_{wet}	= 1217 ft^2

A.2 Flight Conditions for Aerodynamic and S&C Calculations

The SBJ is in quasi-steady level flight at the following flight conditions:

$h = 30{,}000$ ft	$\rho = .000889$ slug/ft^3	$a = 995$ ft/s^2
$V = 597$ ft/s	$M = .60$	$\gamma = 0.0$ deg
$\bar{X}_{cg} = .300$	$m = 342$ slugs	$I_{yy} = 18{,}000$ slug ft^2
$\bar{q} = 158$ lb/ft^2	$\bar{C}_{D_0} = .023$	$K = .073$
$C_L = .299$	$C_D = .0295$	$C_T = .0295$

A.3 Aerodynamics

Given: Flight conditions of Sec. A.2.

Wing aerodynamics

Given:

Airfoil NACA 64-109 (M=0):

$\alpha_0 = -0.5$ deg

$c_{l\alpha} = 0.110$ 1/deg

$ac/c = 0.258$

$c_{m_{ac}} = -0.0175$

Calculated:

$\alpha_{0L} = -0.5$ deg

$$\kappa = .940$$
$$C_{L_\alpha} = 4.67$$
$$C_{m_{ac}} = -0.0175$$

Wing-body aerodynamics

Calculated:
$$\alpha_{0L} = -1.5 \text{ deg}$$
$$C_{L_\alpha} = 4.67$$
$$C_{m_{ac}} = -0.0175$$

Horizontal tail aerodynamics

Given:
Airfoil NACA 64 008 (M=0):
$$\alpha_0 = 0.0 \text{ deg}$$
$$c_{l_\alpha} = 0.110 \text{ 1/deg}$$
$$ac/c = 0.260$$
$$c_{m_{ac}} = 0.0$$
Calculated:
$$\alpha_{0L} = 0.0 \text{ deg}$$
$$\kappa = 0.947$$
$$M_H = 0.569$$
$$C_{L_\alpha} = 4.03$$
$$C_{m_{ac}} = 0.0$$
$$\tau_E = 0.509$$

Airplane aerodynamics

Calculated:
$$\varepsilon_\alpha = 0.420$$
$$\varepsilon_0 = 0.0110$$
$$C_{L_0} = 0.0835$$
$$C_{L_\alpha} = 5.16$$
$$C_{L_{\delta_E}} = 0.430$$
$$\bar{X}_{ac_{WB}} = 0.258$$

$$\bar{X}_{ac_H} = 2.93$$
$$\bar{V}_H = 0.612$$
$$C_{m_0}^A = 0.0895$$
$$C_{m_\alpha}^A = -1.09$$
$$C_{m_{\delta_E}}^A = -1.13$$
$$\bar{X}_{ac} = 0.512$$
$$C_{L_q} = 4.44$$
$$C_{m_q}^A = -11.7$$
$$C_{L_{\dot{\alpha}}} = 1.89$$
$$C_{m_{\dot{\alpha}}}^A = -4.98$$

A.4 Static Longitudinal S&C, Trim Conditions

Given: Flight conditions of Sec. A.2 and the results of Sec. A.3

$$\bar{C}_{D_0} = .0230$$
$$\bar{K} = .0730$$
$$C_L = .299$$
$$C_D = .0295$$
$$C_T = .0295$$
$$T = 1080 \text{ lb}$$
$$C_{m_0}^T = -.0084$$
$$C_{m_0} = .0811$$
$$C_{m_\alpha} = -1.09$$
$$C_{m_{\delta_E}} = -1.13$$
$$\alpha = .0389 \ (2.23 \text{ deg})$$
$$\delta_E = .0341 \ (1.95 \text{ deg})$$

A.5   Dynamic Longitudinal S&C

Given: Flight conditions of Sec. A.2 and results of Secs. A.3 and A.4

Reference conditions

$$C_{L_1} = .299 \qquad C_{D_1} = .0295 \qquad C_{T_1} = .0295$$
$$\Theta_1 = 0.0 \text{ deg} \qquad \bar{\alpha}_1 = 2.23 \text{ deg} \qquad \bar{\delta}_{E_1} = 1.95 \text{ deg}$$
$$C_{x_1} = 0 \qquad C_{z_1} = -.299 \qquad C_{m_1} = 0$$
$$U_1 = 597 \text{ ft/s}$$

Nondimensional and dimensional stability derivatives

C_{D_α}	$= .214$	C_{x_α}	$= .0848$	X_α	$= 9.13 \text{ ft/s}^2$
C_{L_α}	$= 5.16$	C_{x_u}	$= -.0626$	X_u	$= -.0113$
$C_{m_\alpha}^A$	$= -1.09$	$C_{x_{\dot\alpha}}$	$= 0.0$	$X_{\dot\alpha}$	$= 0.0 \text{ ft/s}$
$C_{m_\alpha}^T$	$= 0.0$	C_{x_q}	$= 0.0$	X_q	$= 0.0 \text{ ft/s}$
C_{T_u}	$= -.0591$	$C_{x_{\delta_E}}$	$= 0.0$	X_{δ_E}	$= 0.0 \text{ ft/s}^2$
C_{D_u}	$= .0035$	C_{z_α}	$= -5.19$	Z_α	$= -558 \text{ ft/s}^2$
C_{L_u}	$= .0881$	C_{z_u}	$= -.0881$	Z_u	$= -.124 \text{ 1/s}$
$C_{m_u}^A$	$= .0261$	$C_{z_{\dot\alpha}}$	$= -1.89$	$Z_{\dot\alpha}$	$= -1.19 \text{ ft/s}$
$C_{m_u}^T$	$= .0169$	C_{z_q}	$= -4.38$	Z_q	$= -2.80 \text{ ft/s}$
$C_{D_{\dot\alpha}}$	$= 0.0$	$C_{z_{\delta_E}}$	$= -.430$	Z_{δ_E}	$= -46.2 \text{ ft/s}^2$
$C_{L_{\dot\alpha}}$	$= 1.89$	C_{m_α}	$= -1.09$	M_α	$= -15.6 \text{ 1/s}^2$
$C_{m_{\dot\alpha}}^A$	$= -4.98$	C_{m_u}	$= .0430$	M_u	$= .00100 \text{ 1/(ft s)}$
C_{D_q}	$= 0.0$	$C_{m_{\dot\alpha}}$	$= -4.98$	$M_{\dot\alpha}$	$= -.418 \text{ 1/s}$
C_{L_q}	$= 4.44$	C_{m_q}	$= -11.7$	M_q	$= -.979 \text{ 1/s}$
$C_{m_q}^A$	$= -11.7$	$C_{m_{\delta_E}}$	$= -1.13$	M_{δ_E}	$= -16.2 \text{ 1/s}^2$
$C_{D_{\delta_E}}$	$= 0.0$				
$C_{L_{\delta_E}}$	$= .430$				
$C_{m_{\delta_E}}^A$	$= -1.13$				

Appendix B

Reference Conditions and Stability Derivatives

In this appendix, formulas are given for predicting the reference force and moment coefficients, as well as for the u derivatives. The relationship between the stability derivatives defined in Chap. 11 and those defined in Chap. 8 is established.

Reference Conditions

The reference flight condition is defined by h_1, M_1, m_1g, T_1, and values of α_1 and δ_{E_1} which are obtained from Sec. 9.2. The reference force and moment coefficients are $C_{T_1}, C_{D_1}, C_{L_1}$, and $C_{x_1}, C_{z_1}, C_{m_1}$. The lift coefficient is obtained from Eq. (11.11), that is,

$$C_{L_1} = \frac{m_1 g}{\bar{q}_1 S}. \tag{B.1}$$

Next, for a parabolic drag polar with constant coefficients,

$$C_{D_1} = \bar{C}_{D_0} + \bar{K} C_{L_1}^2 \tag{B.2}$$

Also,

$$C_{T_1} = \frac{T_1}{\bar{q}_1 S}. \tag{B.3}$$

Finally, from Eq. (11.19)

$$C_{x_1} = C_{T_1} - C_{D_1}, \quad C_{z_1} = -C_{L_1}, \quad C_{m_1} = 0. \tag{B.4}$$

α Derivatives

The α derivatives are C_{L_α}, C_{m_α}, and C_{D_α}. Recall that α is the angle of attack of the x_s axis, that $\bar\alpha$ is the angle of attack of the x_b axis, and that the two are related by $\bar\alpha = \bar\alpha_1 + \alpha$. To use the results of Chap. 8, α there must be replaced by $\bar\alpha$.

By definition,

$$C_{L_\alpha} = \frac{\partial C_L}{\partial \alpha}\bigg|_1 = \frac{\partial C_L}{\partial \bar\alpha}\bigg|_1. \tag{B.5}$$

Hence,

$$C_{L_\alpha} = C_{L_{\bar\alpha}}(M_1) \tag{B.6}$$

whose value is given by Eq. (8.46). The same is true for $C_{m_\alpha}^A$, that is,

$$C_{m_\alpha}^A = C_{m_{\bar\alpha}}^A(M_1) \tag{B.7}$$

whose value is obtained from Eq. (8.55) Finally, consider C_{D_α} which is defined as

$$C_{D_\alpha} = \frac{\partial C_D}{\partial \alpha}\bigg|_1 = \frac{\partial C_D}{\partial \bar\alpha}\bigg|_1. \tag{B.8}$$

Hence,

$$C_{D_\alpha} = C_{D_{\bar\alpha}}(M_1) \tag{B.9}$$

From Eq. (8.80), it is seen that

$$C_{D_{\bar\alpha}}(M_1) = C_1(M_1) + 2C_2(M_1)\bar\alpha_1. \tag{B.10}$$

δ_E Derivatives

The δ_E derivatives are $C_{D_{\delta_E}}$, $C_{L_{\delta_E}}$, and $C_{M_{\delta_E}}$. Recall that in this chapter δ_E is the elevator angle perturbation and that $\bar\delta_e$ is the actual elevator angle. The two are related by $\bar\delta_E = \bar\delta_{E_1} + \delta_E$. To use the results of Chap. 8, δ_E there must be replaced by $\bar\delta_E$.

For subsonic airplanes, C_D is not affected by δ_E so that

$$C_{D_{\delta_E}} = 0. \tag{B.11}$$

Next,

$$C_{L_{\delta_E}} = \frac{\partial C_L}{\partial \delta_E}\bigg|_1 = \frac{\partial C_L}{\partial \bar\delta_E}\bigg|_1 \tag{B.12}$$

Hence,

$$C_{L_{\delta_E}} = C_{L_{\bar{\delta}_E}}(M_1) \tag{B.13}$$

whose value is given by Eq. (8.47). The same is true for $C_{m_{\delta_E}}$, that is,

$$C_{m_{\delta_E}}^A = C_{m_{\bar{\delta}_E}}^A(M_1) \tag{B.14}$$

whose value is given by Eq. (8.56).

u Derivatives

The u derivatives arise because the force and moment coefficients are functions of the Mach number and, hence, the velocity. For C_L, it is seen that

$$C_{L_u} = \left.\frac{\partial C_L}{\partial \frac{u}{U_1}}\right|_1 = \left.\frac{\partial C_L}{\partial M}\frac{\partial M}{\partial V}\frac{\partial V}{\partial \frac{u}{U_1}}\right|_1. \tag{B.15}$$

Then, from the definition of the Mach number ($M = V/a$) and $V = U_1 + u$, this derivative becomes

$$C_{L_u} = \left.\frac{\partial C_L}{\partial M}\right|_1 M_1 \tag{B.16}$$

Similar results hold for C_{m_u}, C_{D_u}, and C_{T_u} that is,

$$C_{m_u} = \left.\frac{\partial C_m}{\partial M}\right|_1 M_1, \quad C_{D_u} = \left.\frac{\partial C_D}{\partial M}\right|_1 M_1, \quad C_{T_u} = \left.\frac{\partial C_T}{\partial M}\right|_1 M_1. \tag{B.17}$$

To discuss the M derivatives, consider first the C_L derivative. It is known from Chap. 8 that

$$\begin{aligned} C_L &= C_{L_0}(M) + C_{L_\alpha}(M)\bar{\alpha} + C_{L_{\delta_E}}(M)\bar{\delta}_E \\ &+ C_{L_Q}(M)(\bar{c}Q/2V) + C_{L_{\dot\alpha}}(M)(\bar{c}\dot\alpha/2V). \end{aligned} \tag{B.18}$$

On the reference path, Q and $\dot\alpha$ are zero. Hence,

$$\left.\frac{\partial C_L}{\partial M}\right|_1 = \left.\frac{\partial C_{L_0}}{\partial M}\right|_1 + \left.\frac{\partial C_{L_\alpha}}{\partial M}\right|_1 \bar\alpha_1 + \left.\frac{\partial C_{L_{\delta_E}}}{\partial M}\right|_1 \bar\delta_{E_1}. \tag{B.19}$$

From Sec. 8.7, it seen that

$$\frac{\partial C_{L_0}}{\partial M} = \frac{\partial C_{L_{\alpha W}}}{\partial M}(i_W - \alpha_{0Lw}) + \frac{\partial C_{L_{\alpha H}}}{\partial M}(i_H - \varepsilon_0)\eta_H\frac{S_H}{S} \quad \text{(B.20)}$$
$$- C_{L_{\alpha H}}\frac{\partial \varepsilon_0}{\partial M}\eta_H\frac{S_H}{S}$$

$$\frac{\partial C_{L_\alpha}}{\partial M} = \frac{\partial C_{L_{\alpha W}}}{\partial M} + \frac{\partial C_{L_{\alpha H}}}{\partial M}(1 - \varepsilon_\alpha)\eta_H\frac{S_H}{S} \quad \text{(B.21)}$$
$$- C_{L_{\alpha H}}\frac{\partial \varepsilon_\alpha}{\partial M}\eta_H\frac{S_H}{S}$$

$$\frac{\partial C_{L_{\delta_E}}}{\partial M} = \frac{\partial C_{L_{\alpha H}}}{\partial M}\tau_E\eta_H\frac{S_H}{S}. \quad \text{(B.22)}$$

For the wing or the horizontal tail (Sec. 3.5), it is known that

$$C_{L_\alpha} = \frac{\pi A}{1 + \sqrt{1 + (A/2\kappa)^2[1 + \tan^2 \Lambda_{hc} - M^2]}} \quad \text{(B.23)}$$

Hence,

$$\frac{\partial C_{L_\alpha}}{\partial M} = \frac{C_{L_\alpha}^2(A/2\kappa)^2 M}{\pi A\sqrt{1 + (A/2\kappa)^2[1 + \tan^2 \Lambda_{hc} - M^2]}} \quad \text{(B.24)}$$

For the wing, it is seen that

$$\frac{\partial C_{L_{\alpha W}}}{\partial M} = \frac{C_{L_{\alpha W}}^2(A_W\kappa_W)^2 M}{\pi A_W\sqrt{1 + (A_W/2\kappa_W)^2[1 + \tan^2 \Lambda_{hcw} - M^2]}}. \quad \text{(B.25)}$$

The same result for the horizontal tail is given by

$$\frac{\partial C_{L_{\alpha H}}}{\partial M} = \frac{\partial C_{L_{\alpha H}}}{\partial M_H}\frac{\partial M_H}{\partial M} = \sqrt{\eta_H}\frac{\partial C_{L_{\alpha H}}}{\partial M_H}. \quad \text{(B.26)}$$

where

$$\frac{\partial C_{L_{\alpha H}}}{\partial M} = \frac{C_{L_{\alpha H}}^2(A_H\kappa_H)^2 M}{\pi A_H\sqrt{1 + (A_H/2\kappa_H)^2[1 + \tan^2 \Lambda_{hc_H} - M^2]}}. \quad \text{(B.27)}$$

Finally, the Mach number derivatives of ε_α and ε_0 are obtained from Sec. 8.5 as

$$\frac{\partial \varepsilon_\alpha}{\partial M} = \frac{(\varepsilon_\alpha)_{M=0}}{(C_{L_\alpha})_{M=0}}\frac{\partial C_{L_{\alpha W}}}{\partial M}, \quad \frac{\partial \varepsilon_0}{\partial M} = \frac{\partial \varepsilon_\alpha}{\partial M}(i_W - \alpha_{0Lw}) \quad \text{(B.28)}$$

Eqs. (B.20) through (B.28) lead to $\partial C_{L_0}/\partial M$, $\partial C_{L_\alpha}/\partial M$, and $\partial C_{L_{\delta_E}}/\partial M$ and, hence, $\partial C_L/\partial M$.

For the derivative $\partial C_m/\partial M$, it is recalled from Sec. 8.2 that

$$
\begin{aligned}
C_m &= C_{m_0}^T(M) + C_{m_0}^A(M) + C_{m_\alpha}^A(M)\alpha + C_{m_{\delta_E}}^A(M)\delta_E \\
&+ C_{m_Q}^A(M)(\bar{c}Q/2V) + C_{m_{\dot\alpha}}^A(M)(\bar{c}\dot\alpha/2V)
\end{aligned}
\tag{B.29}
$$

where Q and $\dot\alpha$ are zero on the reference path. Hence,

$$
\left.\frac{\partial C_m}{\partial M}\right|_1 = \left.\frac{\partial C_{m_0}^T}{\partial M}\right|_1 + \left.\frac{\partial C_{m_0}^A}{\partial M}\right|_1 + \left.\frac{\partial C_{m_\alpha}^A}{\partial M}\right|_1 \bar\alpha_1 + \left.\frac{\partial C_{m_{\delta_E}}^A}{\partial M}\right|_1 \bar\delta_{E_1}.
\tag{B.30}
$$

While $\partial C_{m_0}^T/\partial M$ is discussed in the final paragraph, the remaining derivatives are calculated in the same manner as the C_L derivatives. Here,

$$
\begin{aligned}
\frac{\partial C_{m_0}^A}{\partial M} &= \frac{\partial C_{L_{\alpha W}}}{\partial M}(i_W - \alpha_{0LW})(\bar{X}_{cg} - \bar{X}_{ac}) \\
&- \frac{\partial C_{L_{\alpha H}}}{\partial M}(i_H - \varepsilon_0)\eta_H\bar{V}_H + C_{L_{\alpha H}}\frac{\partial\varepsilon_0}{\partial M}\eta_H\bar{V}_H
\end{aligned}
\tag{B.31}
$$

$$
\begin{aligned}
\frac{\partial C_{m_\alpha}^A}{\partial M} &= \frac{\partial C_{L_{\alpha W}}}{\partial M}(\bar{X}_{cg} - \bar{X}_{ac}) - \frac{\partial C_{L_{\alpha H}}}{\partial M}(1 - \varepsilon_\alpha)\eta_H\bar{V}_H \\
&+ C_{L_{\alpha H}}\frac{\partial\varepsilon_\alpha}{\partial M}\eta_H\bar{V}_H
\end{aligned}
\tag{B.32}
$$

$$
\frac{\partial C_{m_{\delta_E}}^A}{\partial M} = -\frac{\partial C_{L_{\alpha H}}}{\partial M}\tau_E\eta_H\bar{V}_H.
\tag{B.33}
$$

The derivative $\partial C_D/\partial M$ is obtained from Sec. 8.11 as

$$
\frac{\partial C_D}{\partial M} = \frac{\partial C_0}{\partial M} + \frac{\partial C_1}{\partial M}\bar\alpha + \frac{\partial C_2}{\partial M}\bar\alpha^2
\tag{B.34}
$$

where

$$
\begin{aligned}
\frac{\partial C_0}{\partial M} &= 2\bar{K}C_{L_0}\frac{\partial C_{L_0}}{\partial M} \\
\frac{\partial C_1}{\partial M} &= 2\bar{K}\frac{\partial C_{L_0}}{\partial M}C_{L_\alpha} + 2\bar{K}C_{L_0}\frac{\partial C_{L_\alpha}}{\partial M} \\
\frac{\partial C_2}{\partial M} &= 2\bar{K}C_{L_\alpha}\frac{\partial C_{L_\alpha}}{\partial M}.
\end{aligned}
\tag{B.35}
$$

The last step is to compute the C_T derivatives. By definition

$$
C_T = \frac{T(h.V.P)}{(1/2)\rho V^2 S} = \frac{T}{(kp/2)M^2 S}, \quad C_{m_0}^T = C_T\frac{l_T}{\bar{c}}.
\tag{B.36}
$$

If it is assumed that atmospheric properties are constant, the thrust is independent of the velocity, and the power setting is constant, the thrust is constant. Hence,

$$\frac{\partial C_T}{\partial M}\bigg|_1 = -\frac{2C_{T_1}}{M_1} \tag{B.37}$$

and

$$\frac{\partial C_{m_0}^T}{\partial M}\bigg|_1 = -\frac{2C_{T_1}l_T}{M_1\bar{c}}. \tag{B.38}$$

q Derivatives

The effect of q on the drag coefficient is negligible so that

$$C_{D_q} = 0. \tag{B.39}$$

By definition,

$$C_{L_q} = \frac{\partial C_L}{\partial \frac{\bar{c}q}{2U_1}}\bigg|_1 = \frac{\partial C_L}{\partial \frac{\bar{c}Q}{2U_1}}\bigg|_1 \tag{B.40}$$

Hence,

$$C_{L_q} = C_{L_Q}(M_1) \tag{B.41}$$

whose value can be obtained from Eq. (8.67). Similarly,

$$C_{m_q}^A = C_{m_Q}^A(M_1), \tag{B.42}$$

whose value is given by Eq. (8.69).

$\dot{\alpha}$ Derivatives

It is assumed that the effect of $\dot{\alpha}$ on the drag coefficient is negligible so that

$$C_{D_{\dot{\alpha}}} = 0. \tag{B.43}$$

The definition of $C_{L_{\dot{\alpha}}}$, that is,

$$C_{L_{\dot{\alpha}}} = \frac{\partial C_L}{\partial \frac{\bar{c}\dot{\alpha}}{2U_1}}\bigg|_1 = \frac{\partial C_L}{\partial \frac{\bar{c}\dot{\alpha}}{2U_1}}\bigg|_1 \tag{B.44}$$

gives

$$C_{L_{\dot{\alpha}}} = C_{L_{\dot{\alpha}}}(M_1) \tag{B.45}$$

whose value can be obtained from Eq. (8.76). Similarly,

$$C_{m_{\dot{\alpha}}}^{A} = C_{m_{\dot{\alpha}}}^{A}(M_1). \tag{B.46}$$

Its value is given by Eq. (8.78)

Appendix C

Elements of Linear System Theory

No matter what order a constant-coefficient linear system may be, its response is the sum of the responses of first-order and second-order systems. Hence, first-order and second-order systems are the subject of this appendix. First, Laplace transforms are presented. Laplace transforms are used because they convert linear ordinary differential equations with constant coefficients into algebraic polynomial equations which are easily solved. Also, it is possible to carry out the design of single-input single-output systems in the frequency domain, that is, in the s-plane resulting from the Laplace transform. Next, first-order systems are investigated. The response to a step input and the stability of the response are topics of interest. Finally, the same analysis is performed for a second-order system.

C.1 Laplace Transforms

The Laplace transform is formally defined as

$$x(s) = \int_0^\infty e^{-st} x(t) dt \equiv L\{x(t)\} \qquad \text{(C.1)}$$

where $t > 0$ and s is a complex variable. The complete derivation of a Laplace transform is not a trivial matter; however, it is possible to accomplish most work by using the transforms listed below.

$$L\{cx(t)\} = cL\{x(t)\}, \quad c \equiv \text{constant} \qquad \text{(C.2)}$$

$$L\{1\} = 1/s \tag{C.3}$$
$$L\{t\} = 1/s^2 \tag{C.4}$$
$$L\{t^{n-1}/(n-1)!\} = 1/s^n \tag{C.5}$$
$$L\{e^{at}\} = 1/(s-a) \tag{C.6}$$
$$L\{\delta(t)\} = 1, \quad \delta(t) \equiv \text{unit impulse} \tag{C.7}$$
$$L\{e^{at}x(t)\} = x(s-a) \tag{C.8}$$
$$L\left\{\frac{dx}{dt}\right\} = -x(0) + sx(s), \quad x(0) = x(t) \text{ at } t = 0 \tag{C.9}$$
$$L\left\{\frac{d^2x}{dt^2}\right\} = -\dot{x}(0) - sx(0) + s^2x(s) \tag{C.10}$$
$$L\{\sin \omega t\} = \frac{\omega}{s^2 + \omega^2} \tag{C.11}$$
$$L\{\cos \omega t\} = \frac{s}{s^2 + \omega^2} \tag{C.12}$$

The reverse process, s to t, is accomplished by working backwards in the table.

C.2   First-Order System

The standard form of a first-order system is given by

$$\tau\dot{x}(t) + x(t) = u(t) \tag{C.13}$$

where τ is a constant and u is the input. If zero initial conditions are assumed, the Laplace transform of this equation leads to

$$\tau s x(s) + x(s) = u(s) \tag{C.14}$$

which can be rewritten in the form of output over input as

$$\frac{x(s)}{u(s)} = \frac{1}{\tau s + 1}. \tag{C.15}$$

The right-hand side is called the *transfer function* of the system, that is, the ratio of the output to the input.

Response to a step input

For a step input, u has the form

$$u(t) = c \tag{C.16}$$

where c is a constant. The corresponding Laplace transform is

$$u(s) = c/s \qquad (C.17)$$

from Eqs. (C.2) and (C.3). Then, the output is obtained from Eq. (C.15) as

$$x(s) = \frac{c}{s(\tau s + 1)}. \qquad (C.18)$$

To find the response in the time domain, apply the method of partial fractions, that is, assume

$$x(s) = \frac{A_1}{s} + \frac{A_2}{\tau s + 1} \qquad (C.19)$$

and note that

$$A_1 = \lim_{s \to 0} sx(s) = \lim_{s \to 0} \frac{sc}{s(\tau s + 1)} = c \qquad (C.20)$$

$$A_2 = \lim_{s \to -\frac{1}{\tau}} (\tau s + 1)x(s) = \lim_{s \to -\frac{1}{\tau}} \frac{(\tau s + 1)c}{s(\tau s + 1)} = -\tau c. \qquad (C.21)$$

Hence,

$$x(s) = \frac{c}{s} - \frac{\tau c}{\tau s + 1} = \frac{c}{s} - \frac{c}{s + \frac{1}{\tau}} \qquad (C.22)$$

and, from the Laplace transform formulas, the time response is

$$x(t) = c - ce^{-\frac{1}{\tau}t} = c(1 - e^{-\frac{t}{\tau}}). \qquad (C.23)$$

Note that the time constant $T = -1/\tau$ is the time required to reach 64% of the steady state value.

The nature of the solution depends on the poles of the transfer function which are the roots of the characteristic equation $\tau s + 1 = 0$. There are three cases: (1) $\tau > 0$, (2) $\tau = 0$, and (3) $\tau < 0$.

To prevent the response from increasing with time, it is necessary that $\tau > 0$ where τ is called the time constant of the response. In the s-plane, where s is the complex number $s = n + i\omega$, this means that the pole must lie in the left half plane as shown in Fig. C.1. The response (C.23) also is shown in Fig. C.1. Such a system is said to be dynamically stable because the transient part of the response (the term involving time) goes to zero.

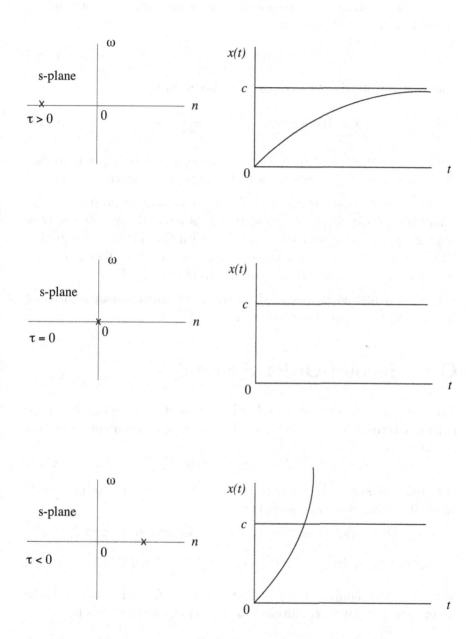

Figure C.1: Response vs. Pole Location - First-Order System

The steady state response of a stable system is obtained by letting $t \to \infty$ and is given by

$$x(\infty) = c. \tag{C.24}$$

Note that the steady state response is also given by

$$x(\infty) = \lim_{s \to 0} sx(s) = \lim_{s \to 0} s\frac{c}{s(\tau s + 1)} = c \tag{C.25}$$

which is known as the final-value theorem. It is interesting to note that the output $x(t)$ of the system tracks the input $u = c$ since $x(t) \to c$.

For the case where $\tau = 0$, the time dependent term in Eq. (C.23) has an exponent of $-\infty$ so that it is zero. Hence, the response is given by $x(t) = c$ which is consistent with the differential equation (C.13). Such a response is neutrally stable because it neither grows nor decays with time. These results are illustrated in Fig. C.1.

Finally, when $\tau < 0$, the response (C.23) is unstable because the exponential term grows with time. See Fig. C.1.

C.3   Second-Order System

There are two forms of the second-order system depending on whether or not the x term is present. Without the x term, the second-order system is written as

$$\tau \ddot{x}(t) + \dot{x}(t) = u(t). \tag{C.26}$$

Note that this system is first-order in $\dot{x}$, and its solution can be integrated to obtain x, which is non-oscillatory.

The standard form of a second-order system is given by

$$\ddot{x}(t) + 2\zeta\omega_n\,\dot{x}(t) + \omega_n^2 x(t) = u(t) \qquad \zeta, \omega_n = \text{consts} \geq 0 \tag{C.27}$$

where ζ is the damping ratio, ω_n is the natural frequency, and u is the input. For zero initial conditions, Laplace transforming leads to

$$s^2 x(s) + 2\zeta\omega_n sx(s) + \omega_n^2 x(s) = u(s) \tag{C.28}$$

so that

$$\frac{x(s)}{u(s)} = \frac{1}{s^2 + 2\zeta\omega_n s + \omega_n^2} \tag{C.29}$$

where the right-hand side is the transfer function.

Response to a step input

For the step input (C.16) and its Laplace transform (C.17), the output of the second-order system is given by

$$x(s) = \frac{c}{s(s^2 + 2\zeta\omega_n s + \omega_n^2)} = \frac{c}{s(s - \lambda_1)(s - \lambda_2)} \tag{C.30}$$

where

$$\lambda_1 = -\omega_n\zeta + \omega_n\sqrt{\zeta^2 - 1} \tag{C.31}$$

$$\lambda_2 = -\omega_n\zeta - \omega_n\sqrt{\zeta^2 - 1}. \tag{C.32}$$

The process for finding the response in the time domain, that is, the method of partial fractions, depends on whether the roots λ_1 and λ_2 of the characteristic equation $s^2 + 2\zeta\omega_n s + \omega_n^2 = 0$ are distinct or equal.

If the roots are distinct ($\lambda_1 \neq \lambda_2$), the method of partial fractions leads to

$$x(s) = \frac{A_1}{s} + \frac{A_2}{s - \lambda_1} + \frac{A_3}{s - \lambda_2} \tag{C.33}$$

where

$$A_1 = \lim_{s \to 0} sx(s) = \frac{c}{\lambda_1\lambda_2} \tag{C.34}$$

$$A_2 = \lim_{s \to \lambda_1} (s - \lambda_1)x(s) = \frac{c}{\lambda_1(\lambda_1 - \lambda_2)} \tag{C.35}$$

$$A_3 = \lim_{s \to \lambda_2} (s - \lambda_2)x(s) = \frac{c}{\lambda_2(\lambda_2 - \lambda_1)}. \tag{C.36}$$

Hence, the frequency domain response is given by

$$x(s) = \frac{c}{\lambda_1\lambda_2}\frac{1}{s} + \frac{c}{\lambda_1(\lambda_1 - \lambda_2)}\frac{1}{s - \lambda_1} + \frac{c}{\lambda_2(\lambda_2 - \lambda_1)}\frac{1}{s - \lambda_2} \tag{C.37}$$

so that the time domain response becomes

$$x(t) = \frac{c}{\lambda_1\lambda_2} + \frac{c}{\lambda_1(\lambda_1 - \lambda_2)}e^{\lambda_1 t} + \frac{c}{\lambda_2(\lambda_2 - \lambda_1)}e^{\lambda_2 t}. \tag{C.38}$$

If the roots of the characteristic equations are equal ($\lambda_2 = \lambda_1$) or repeated, the method of partial fractions changes somewhat. Here, the transfer function becomes

$$x(s) = \frac{c}{s(s - \lambda_1)^2} \tag{C.39}$$

and the partial fractions decomposition is written as

$$x(s) = \frac{A_1}{s} + \frac{B_1}{s - \lambda_1} + \frac{B_2}{(s - \lambda_1)^2} = \frac{A_1}{s} + \frac{B_1(s - \lambda_1) + B_2}{(s - \lambda_1)^2}. \quad (C.40)$$

As before,

$$A_1 = \lim_{s \to 0} s x(s). \quad (C.41)$$

However, for the repeated roots,

$$B_2 = \lim_{s \to \lambda_1} (s - \lambda_1)^2 x(s) \quad (C.42)$$

$$B_1 = \lim_{s \to \lambda_1} \frac{d}{ds}[(s - \lambda_1)^2 x(s)]. \quad (C.43)$$

Hence,

$$A_1 = \frac{c}{\lambda_1^2} \quad (C.44)$$

$$B_1 = -\frac{c}{\lambda_1^2} \quad (C.45)$$

$$B_2 = \frac{c}{\lambda_1} \quad (C.46)$$

and

$$x(s) = \frac{c}{\lambda_1^2 s} - \frac{c}{\lambda_1^2 (s - \lambda_1)} + \frac{c}{\lambda_1 (s - \lambda_1)^2}. \quad (C.47)$$

In the time domain,

$$x(t) = \frac{c}{\lambda_1^2} - \frac{c}{s_1^2} e^{\lambda_1 t} + \frac{c}{\lambda_1} t e^{\lambda_1 t}. \quad (C.48)$$

The specific response of the system depends on amount of damping ζ in the system. There four cases: (1) $\zeta > 1$, (2) $\zeta = 1$, (3) $1 > \zeta > 0$, and (4) $\zeta = 0$. These cases are discussed separately below.

Case 1: $\zeta > 1$

In this case, the poles λ_1 and λ_2 given by Eqs. (C.31) and (C.32) are real and distinct. Since $\zeta > 1$, λ_1 and λ_2 are both negative and lie on the negative real axis of the complex s-plane: (see Fig. C.2). As $\zeta \to \infty$, λ_1 tends to the origin ($\lambda_1 \to 0$) and λ_2 tends to negative infinity ($\lambda_2 \to -\infty$). Note that since $\lambda_1 < 0$ and $\lambda_2 < 0$, the response is stable in the sense that the time terms in (C.38) die out (tend to zero), and the system goes into steady state. The response in the time domain is shown in Fig. C.2. The steady-state response is obtained from Eq. (C.38) by letting $t \to \infty$ and is given by

$$x(\infty) = \frac{c}{\lambda_1 \lambda_2} = \frac{c}{\omega_n^2}. \tag{C.49}$$

Note that the output differs from the input by the factor $1/\omega_n^2$, so that, the system does not track the input.

The steady-state output can also be obtained by applying the final value theorem

$$x(\infty) = \lim_{s \to 0} sx(s) = s\frac{c}{s(s^2 + 2\zeta\omega_n s + \omega_n^2)} = \frac{c}{\omega_n^2}. \tag{C.50}$$

This result checks with the time-domain result (C.49).

Case 2: $\zeta = 1$

For this case, the roots (C.31) and (C.32) are equal and given by (Fig. C.2)

$$\lambda_1 = \lambda_2 = -\omega_n. \tag{C.51}$$

The response in the time domain [Eq. (C.48)] becomes

$$x(t) = \frac{c}{\omega_n^2} - \frac{c}{\omega_n^2}e^{-\omega_n t} - \frac{c}{\omega_n}te^{-\omega_n t} \tag{C.52}$$

and is shown in Fig. C.2. The response is stable because the transient response (the time terms in Eq. (C.52)) goes to zero. To show this, it is necessary to apply L'Hospital's rule to the third term in the form $t/e^{\omega_n t}$ as $t \to \infty$. Then, the steady-state response becomes

$$x(\infty) = \frac{c}{\omega_n^2} \tag{C.53}$$

meaning that the output does not track the input. This result can also be obtained by applying the final value theorem to the response (C.47), that is

$$x(\infty) = \lim_{s \to 0} sx(s) = \frac{c}{\lambda_1^2} = \frac{c}{\omega_n^2}. \tag{C.54}$$

Case 3: $1 > \zeta > 0$

In this case, the poles (C.31) and (C.32) are imaginary and, in fact, are the complex conjugates

$$\lambda_1 = -\omega_n \zeta + i\omega_n \sqrt{1 - \zeta^2} \tag{C.55}$$

$$\lambda_2 = -\omega_n \zeta - i\omega_n \sqrt{1 - \zeta^2}. \tag{C.56}$$

Hence the poles lie in the left-hand s-plane as shown in Fig. C.2. For $\zeta \to 1$, the poles move toward the real axis, and for $\zeta \to 0$, the poles move towards the imaginary axis.

For imaginary poles which are distinct, the response is oscillatory as shown by writing the roots as

$$\lambda_1 = n + i\omega \tag{C.57}$$

$$\lambda_2 = n - i\omega \tag{C.58}$$

where

$$n = -\omega_n \zeta, \quad \omega = \omega_n \sqrt{1 - \zeta^2}. \tag{C.59}$$

Then, if Euler's formula

$$e^{i\phi} = \cos\phi + i\sin\phi \tag{C.60}$$

is applied, the response in the time domain (C.38) becomes (Fig. C.2)

$$x(t) = \frac{c}{n^2 + \omega^2}\left[1 - e^{nt}\left(\cos\omega t - \frac{n}{\omega}\sin\omega t\right)\right] \tag{C.61}$$

or

$$x(t) = \frac{c}{\omega_n^2}\left[1 - e^{-\omega_n \zeta t}\left(\cos\omega_n\sqrt{1 - \zeta^2}\, t + \frac{\zeta}{\sqrt{1 - \zeta^2}}\sin\omega_n\sqrt{1 - \zeta^2}\, t\,\right)\right] \tag{C.62}$$

or

$$x(t) = \frac{c}{\omega_n^2}\left[1 - \frac{e^{-\omega_n \zeta t}}{\sqrt{1 - \zeta^2}}\sin(\omega_n\sqrt{1 - \zeta^2}t + \phi)\right] \tag{C.63}$$

where $\phi = \sin^{-1}\sqrt{1 - \zeta^2}$ is called the phase angle.

It is easily seen that the response is stable because the exponential term goes to zero as time becomes infinite. Also, the steady-state response is given by

$$x(\infty) = \frac{c}{\omega_n^2} \tag{C.64}$$

a result which can be verified by applying the final-value theorem to the response (C.30).

Case 4: $\zeta = 0$

Here, the poles (C.31) and (C.32) are the complex conjugates

$$\lambda_1 = i\omega_n \tag{C.65}$$
$$\lambda_2 = -i\omega_n, \tag{C.66}$$

but they lie on the imaginary axis (Fig. C.2) because there is no damping. The time response is obtained from Eq. (C.61) by setting $n = 0$ and is given by

$$x(t) = \frac{c}{\omega_n^2}\left[1 - \cos\omega_n t\right] \tag{C.67}$$

and is plotted in Fig. C.2. From this result, it is seen that the response is a pure oscillation. Such a response is neither stable nor unstable and is said to be neutrally stable. Since the time term (transient response) does not vanish as $t \to \infty$, there is no steady-state response. Also, the final value theorem does not apply because the system is not stable.

Remark: In the event the damping is negative ($\zeta < 0$), the transient terms grow with time, and the system is unstable. See Fig. C.2.

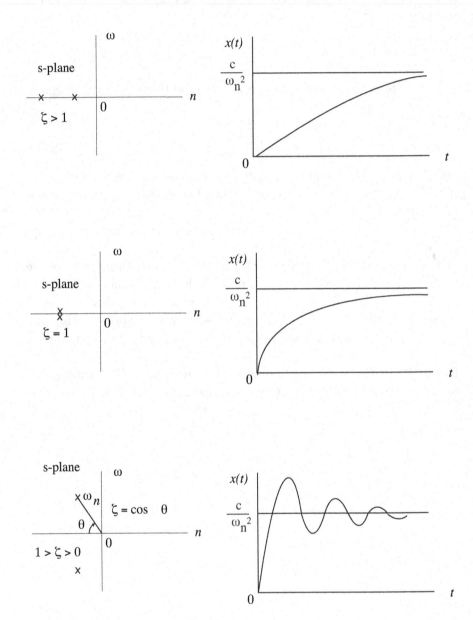

Figure C.2: Response vs. Pole Location - Second-Order System

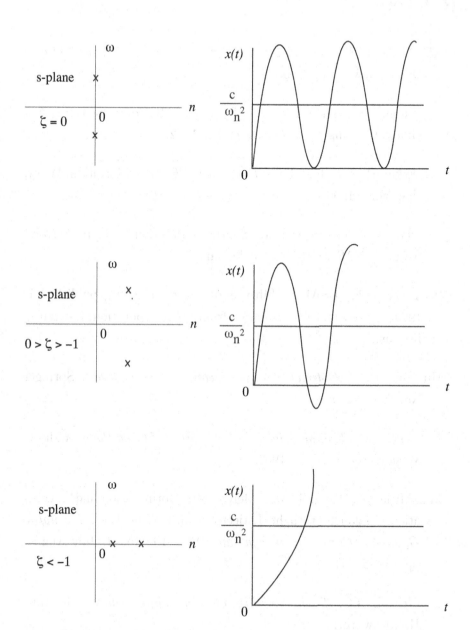

Figure C.2 (cont.): Response vs. Pole Location - Second-Order System

References

AD Abott, I. H., and von Doenhoff, A. E., *Theory of Wing Sections*, Dover, New York, 1959.

An Anon, U.S. Standard Atmosphere, 1962, U.S. Government Printing Office, Washington, D.C., December 1962.

ER Etkin, B., and Reid, L.D., *Dynamics of Flight*, 3nd Edition, Wiley, New York, 1996.

Ga "GASP - General Aviation Synthesis Program, Vol. III, Aerodynamics", NASA CR-152303, January, 1978.

Ho Hoak, D.E., USAF Stability and Control Datcom, published in 1960, revised in 1978, available from DARcorporation, Lawrence, Kansas.

Hu Hull, D.G., *Optimal Control Theory for Applications*, Springer, New York, 2003.

Mi1 Miele, A., *Flight Mechanics, Vol. I, Theory of Flight Paths*, Addison-Wesley, Reading MA, 1962.

Mi2 Miele, A., Wang, T., and Melvin, W., Optimization and Acceleration Guidance of Flight Trajectories in a Windshear, *Journal of Guidance, Control, and Dynamics*, Vol. 10, No. 4, July-August pp. 368-377, 1987.

Ne Nelson, R.C., *Flight Stability and Automatic Control*, McGraw-Hill, New York, 1989.

Pa Pamadi, B.N., *Performance, Stability, Dynamics, and Control of Airplanes*, Second Edition, American Institute of Aeronautics and Astronautics, Reston VA, 2004.

Pe Perkins, Courtland D. *"Development of Airplane Stability and Control Technology"*, AIAA Journal of Aircraft, Vol. 7, No. 4, July-August, 1970.

Ro1 Roskam, J., *Methods for Estimating Stability and Control Derivatives of Conventional Subsonic Airplanes*, DARcorporation, Lawrence, Kansas, 1971.

Ro2 Roskam, J., *Flight Dynamics of Rigid and Elastic Airplanes, Vol. 1*, DARcorporation, Lawrence, Kansas, 1972.

Sc Schemensky, R.T., *Development of an Emperically Based Computer Program to Predict the Aerodynamic Characteristics of Aircraft, Volume 1, Empirical Methods*, Technical Report AFFDL-TR-73-144, Wright-Patterson Air Force Base, Ohio, November 1973.

Ye Yechout, T.R., *Introduction to Aircraft Flight Mechanics*, American Institute of Aeronautics and Astronautics, Reston VA, 2004.

Index